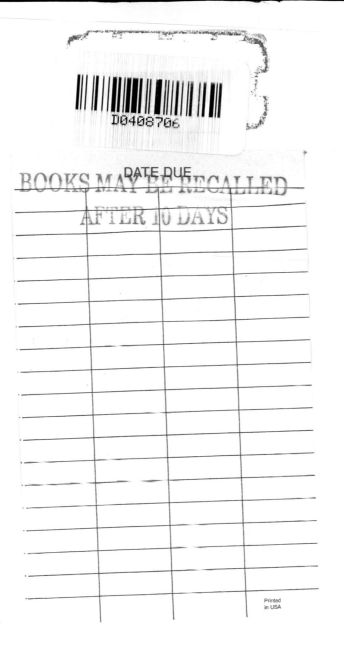

DATE DUE

BOOKS MAY BE RECALLED
AFTER 10 DAYS

Printed
in USA

Chinese Telecommunications Policy

For a listing of recent titles in the *Artech House Telecommunications Library,* turn to the back of this book.

Chinese Telecommunications Policy

Xu Yan
Douglas Pitt

Artech House
Boston • London
www.artechhouse.com

Library of Congress Cataloging-in-Publication Data
Yan, Xu.
    Chinese telecommunications policy / Xu Yan, Douglas Pitt.
        p.   cm. — (Artech House telecommunications library)
    Includes bibliographical references and index.
    ISBN 1-58053-328-0 (alk. paper)
        1. Telecommunication—Government policy—China.   I. Pitt, Douglas C.   II. Title.
    III. Series.
TK5102.3.C49 Y35 2002
384'.0951—dc21   2002074696

British Library Cataloguing in Publication Data
Yan, Xu
    Chinese telecommunications policy.—(Artech House telecommunications library)
    1. Telecommunication policy—China
    I. Title   II. Pitt, Douglas, 1943–
    384' .0951

    ISBN 1-58053-328-0

Cover design by Igor Valdman

© 2002 ARTECH HOUSE, INC.
685 Canton Street
Norwood, MA 02062

International Standard Book Number: 1-58053-328-0
Library of Congress Catalog Card Number: 2002074696

10 9 8 7 6 5 4 3 2 1

Contents

1

Introduction

Contemporary telecommunications is undergoing rapid change. Newly developed technology and huge demands for the creation, transmission, processing, and provision of information have led to great innovations in telecommunications infrastructure. Computer technology is now deeply embedded in the process of telecommunications, which is best summarized as the digital convergence phenomenon. This has brought about not only a revolution in the way that information is transmitted and processed, but also in the creation of a variety of new services that are mainly exemplified in the form of data processing. As a result, the concept of telecommunications itself has been transformed, and such developments have summoned a diverse new constituency of customers and the emergence of an "infocommunication" industry [1]. Inevitably, the traditional regulatory envelope has been strained, and the introduction of a competitive mechanism has been raised as a policy option by governments in many countries. Such bold experiments as the privatization of British Telecom (BT) and the divestiture of AT&T during the Thatcher and Reagan administrations, respectively, led to a fundamental change in telecommunications provisioning (or paradigm shift). This has entailed a complete reassessment of the regulatory rules of the game, thus creating a new era in the telecommunications systems of a constellation of countries.

Few countries have remained impervious to this process. Companies and governments around the world have been made increasingly aware that an advanced infrastructure and prices that are internationally competitive are

1

central to growth and prosperity. The trend towards radical changes in provisioning strategies is readily discernible not only in countries of the so-called first world but also in those of the developing world. Competition and deregulation are no longer empty slogans, but strategic tools that have been applied in many countries.

Individual states have taken different steps according to their respective economic and political backgrounds. It could be argued that all are still under experiment. Bearing in mind that telecommunications deregulation has taken place only since the 1980s and that telecommunications systems in 2002, especially in developing countries, are still far removed from the future information superhighway and its further explosive developments, it is not unreasonable to conclude that the construction of an optimum deregulatory paradigm remains an idealized objective. At present, there does not seem to be one universal best-way approach to deregulation that all countries should accept, as the telecommunications sectors of various countries are subject to different cultural and political influences that configure different policy outcomes. Technologically deterministic arguments (which were in vogue several decades ago) implying that technological change drives the same public policy changes irrespective of country of origin now seem highly suspect. The implications of this observation are that experience in one country might not be a viable model for policy developments elsewhere. For example, the privatization of BT in the United Kingdom arguably has proven to be a great success since it commenced in 1984, but the privatization philosophy still sits uneasily with the communist convictions of the Chinese government, especially in terms of infrastructure industries. This implies that the launch pad for Chinese telecommunications deregulation cannot be based on the same taken-for-granted cultural norms supporting privatization as those that were found in the United Kingdom.

Additionally, competition itself is not the end of telecommunications deregulation, but arguably just a means to an end. Evidence from Britain and Hong Kong reveals that telecommunications deregulation is a long-term process that needs patient negotiations between regulators and operators and fierce bargains struck between operators. The opening of markets does not automatically mean that any operator can enter the market freely and earn the economic benefit from competition without experiencing any cost. For example, O'Malley [2] challenges the neoclassical view of such developments by pointing out three entry barriers encountered by newcomers: (1) economies of scale enjoyed by incumbent operators, (2) difficulties in overcoming the advantages of an established brand name or product line, and (3) the requirement for significant capital expenditures. In addition, there are also some claimed disadvantages of competition. For example, it will likely increase the element of risk in making loans to or investments in the main network operator, thus increasing the cost and slowing down the pace of extension of these benefits to households and to businesses outside the main cities. Additionally, this can prejudice attainment of the goal of universal service. With the existence of these barriers and disadvantages, it is premature to celebrate competitive determinism in telecommunications. As with technology, so with competition. The competitive paradigm will find more fertile soil for growth in one country compared to others. As pointed out by Pitt and Trauth, given the ascendancy of the "competitive idea" in telecommunications over the past two to three decades, "a key question for policy makers and members of this industry is what are the limits to the competitive paradigm" [3].

This book will attempt to outline and examine trends existing in the current Chinese telecommunications regime, and highlight the limits of the telecommunications competitive paradigm in China. The liberalization of the Chinese telecommunications market, which was marked by the establishment of China Unicom, has attracted the interest of many researchers and policy makers throughout the world. Such interest is due not just to the significant size of the Chinese market, but particularly to the nuances arising from the specific political and economic context of Chinese policy making.

In the past two decades, Chinese telecommunications policy has, arguably, experienced a seismic revolution. This revolution began with the incorporation of market mechanisms into the telecommunications sector for the sake of reforming its historically suboptimal administration and operation. This reform was also accompanied by several preferential policies accorded to the telecommunications sector designed as measures to facilitate the expansion of networks and thus enhance the strategic role of a sophisticated information infrastructure in achieving the government's ambition of modernizing China's economy.

The establishment of China Unicom in 1994 marked the termination of the monopoly operation of telecommunications in China. The transition from monopoly to duopoly signaled the beginning of a transition from state planning to market forces in real terms. The establishment of the Ministry of Information Industry (MII) in 1998 has further facilitated this transition process. This quasi-independent regulator has begun to effectively reshape the regulatory framework and, as a consequence, the competitive landscape of the Chinese telecommunications market. When China formally acceded to the World Trade Organization (WTO) in December 2001, there were seven licensees competing in virtually all service segments of the Chinese telecommunications market.

The revolution in telecommunications policy in China has effectively accelerated the development of the telecommunications infrastructure. In 2002, China has the largest mobile communications network and the second largest fixed communications network in the world. Its teledensity (i.e., the total number of fixed telephone mainlines per 100 inhabitants) has risen from 0.38% in 1978 to 13.9% in 2001. This is undoubtedly an accelerated process, as it has normally taken more than 50 years to achieve such progress in most other countries.

China's accession to the WTO marked the beginning of a new era in its telecommunications sector. However, although the end of the 5-year transitional period will almost certainly result in a fully contestable Chinese market, the regulator is likely to nuance that market with Chinese characteristics. Specific historical, political, and economic factors in China have led to a unique policy paradigm within, inter alia, the telecommunications sector, and the inertia of this paradigm will undoubtedly affect adopted regulatory policies in coming years. To understand the potential regulatory approach and regulatory philosophy in China's post-WTO era, it seems that now is a particularly opportune time to conduct a comprehensive review of telecommunications policy in China, especially that of the past two decades. The present study will focus on the following questions:

1. For a country whose essential telecommunications development started barely two decades ago, and in which there is a very low telephone penetration rate (especially in its rural areas), the Chinese deregulation process, in fact, appears to have commenced from a relatively disadvantaged position compared with other deregulated countries. For example, in the United Kingdom the percentage of residential connections per 100 households was 78% in 1984 when British Telecom was privatized; when AT&T was broken up in the same year, the equivalent figure in the United States was actually more than 100% (i.e., there was more than one residential telephone connection per household) [4]. In China the percentage of residential connections per 100 households was only 5.9% when China Unicom was established in 1994. These examples clearly raise the important question of the appropriate stage or time for the take-off into a full-blown process of telecommunications deregulation.

2. The majority ownership of all operators is retained by the state. This contrasts strongly with the generally observed phenomenon in other countries that telecommunications competition typically happens only among privatized operators, and it has been reasonably thought

that privatization is part and parcel of a deregulation process [5]. The key issue is whether privatization is essential for harvesting the full benefits of competition.

3. What is the most appropriate regulatory framework for the Chinese telecommunications market? Can a regulator closely affiliated with the incumbent create a level playing field in a country where the codification of a system of telecommunications law has yet to be formulated? What were the rationales for establishing the MII and conferring upon it regulatory powers?

4. In a socialist country that is undergoing a period of transition from a planned economy to a market structure, there is an important question of how existing telecommunication operators react to competition. Does the incumbent operator create similar barriers to new entrants, as has happened in first-mover countries? If it does, how do those same new entrants fight to overcome these barriers under a regulatory scheme with admittedly Chinese characteristics? In brief, do Chinese operators behave similarly to their counterparts abroad in a fully competitive market, and can both the telecommunications sector itself and customers as a whole achieve the benefits of competition?

5. The MII, which was established in 1998, is currently a regulator for both telecommunications operation and equipment manufacturing. In this case, what, if any, constraints must it take into account when formulating telecommunications policy? For example, is it able to take a technology-neutral stance (e.g., to issue licenses without designating a specific technology), following the practice of the United States Federal Communications Commission (FCC), which is a professional regulatory agency solely overseeing telecommunications operations?

6. For a telecommunications market that has strictly banned foreign direct investment, what are the implications for both operators and investors of China's accession to the WTO? Have the regulatory and legal frameworks and industrial structure been properly calibrated so as to allow for rapid response of the telecommunications industry to the more kinetic competitive world emerging in the post-WTO era?

These pertinent issues provide fertile ground for research in telecommunications policy. The experience of Chinese telecommunications deregulation provides a useful lesson for other developing countries, especially those former socialist countries that are undergoing transition from a centrally planned economy to the adoption of full-fledged market mechanisms.

The purpose of this book is to provide an in-depth analysis of the above questions while at the same time assessing the fate of the telecommunications competition policy paradigm in a late-mover country with clear differences arising from a political and economic system standing in clear contrast to its first-mover counterparts.

The present account is based upon a literature review and fieldwork carried out in China. This fieldwork occurred in two tranches, first in the period of 1994 to 1996 when both authors were resident in Scotland. Later, more intensive fieldwork was conducted since late 1997 when one of the authors relocated to Hong Kong. A selection of senior regulatory officials and a selected group of telecommunications operation executives were targeted and interviewed. Since 2000, one of the authors has participated in several training programs for officials of the MII and executives from Chinese operators, such as China Telecom and China Mobile. This provided the opportunity for face-to-face contact with telecommunications "influentials" and an excellent opportunity to pose key questions openly to leaders of the Chinese telecommunications sector. The authors are, of course, deeply grateful to them for their generous cooperation during interviews and discussions.

Additionally, the book has benefited tremendously from the close cooperation and enduring friendship between the two authors since the end of 1992. A selection of the material in this book is derived from previous publications. Accordingly, the authors would like to acknowledge the assistance of the various editors of and contributors to these publications including Niall Levine, Lihe Liu, Gary Madden, Erik Bohlin, David Loomis, Haiyang Li, Anne Tsui, Chung-Ming Lau, Jacques Arlandis, Lara Srivastava, and Tim Kelly. Special thanks are due to Gong Min and Ye Ru-yi for devoting their time so generously to the checking of all references and the formatting of all chapters. Also, special thanks are due to Mrs. Noni Gordon at UCT for her proofreading skills, micromanagement of corrections, and for injecting sanity into the production process at times of difficulty!

Finally, special thanks are due to Paul Nihoul. It was with his initiative and encouragement that we undertook the task of writing this book. The book editor, Ruth Harris, has offered robust support and enduring patience throughout the entire process: for both we are most grateful. We sincerely appreciate the help of the anonymous reviewer for providing an insightful critique of all chapters. Such comments and criticisms proved extremely helpful to the authors in sharpening the analysis.

Contents of This Book

Following the introduction in Chapter 1, Chapter 2 provides a general review of the development of telecommunications in China before 1994 with particular emphasis on the government's reform schemes during the period from 1978 to 1994. Analysis of the incumbent's behavior during successive stages of development ultimately raises the issue of the necessity of liberalizing the Chinese telecommunications market.

Chapter 3 examines the overall background behind the establishment of China Unicom. Based on detailed analyses of technological change, policy transference, and pressure from large business users, the authors explicitly argue that the condition is ripe for China to liberalize its telecommunications market, even though there exists a huge gap in infrastructural development between China and other early-mover countries.

On the basis of fieldwork carried out in China, Chapter 4 provides a case study of China Unicom. China Unicom's internal strengths and weaknesses, as well as its external environment (i.e., opportunities and threats) are analyzed in detail. The study clearly shows that an unbalanced regulatory framework can easily turn the relatively vulnerable new entrant into a victim of competition, especially when tackling bottlenecks, such as the deployment of network interconnection.

Chapter 5 examines the competitive scenario after the regulatory framework restructuring in 1998. Enhanced regulatory support for the new entrant, additional market liberalization, and the transition to an efficiency-focused strategy on the part of the incumbent clearly indicate that an independent regulatory agency is critical for telecommunications deregulation.

Chapter 6 analyses the impact of the government's strategy of supporting domestic manufacturing industry on its telecommunications policy. In a developing economy displaying clear modernization ambitions, a sector-specific policy in telecommunications is, in fact, a function of multiple variables. In particular, the economic context of third-generation (3G) licensing clearly indicates that telecommunications policy has spillover consequences for other industrial sectors, such as equipment manufacturing.

Chapter 7 reviews the financing schemes of the Chinese telecommunications sector and highlights the importance of attracting foreign investment. It also signals the potential for suboptimal performance in the telecommunications sector in the post-WTO era if current regulatory and legal frameworks remain impervious to further restructuring. Additionally, it points to the urgency of privatizing the Chinese telecommunications system.

Chapter 8 offers some concluding remarks.

References

[1]   Fransman, M., "Evolution of the Telecommunications Industry into the Internet Age," *Communications & Strategies*, No. 43, Third Quarter, 2001, pp. 57–113.

[2]   O'Malley, E., *Industry and Economic Development: The Challenger for the Latecomer*, Dublin, Ireland: Gill and Macmillan, 1989.

[3]   Trauth, E. M., and D. C. Pitt, "Competition in the Telecommunications Industry: A New Global Paradigm and Its Limits," *J. Information Technology*, Vol. 7, No. 1, 1992, pp. 3–11.

[4]   Summerscale, J., and S. Millman, *British Telecom*, London, UK: de Zoete & Bevan, 1984.

[5]   Ulrich, I., *The Objectives of Competition Policy and Its Contribution to Economic Development*, Paris, France: OECD, 1991.

2

From Administered to Market Governance? Telecommunications Development in China Pre-1994

In China, the development of telecommunications before the market was liberalized in 1994 can be divided into three stages: (1) foreign dominance before the Communist Party acceded to power in 1949, (2) slow network expansion due to chaotic political disturbance from 1949 to 1978, and (3) drastic reform in enterprise management afterwards. Since knowledge of prior development is critical for understanding the current market liberalization in the Chinese telecommunications sector, this chapter will provide a general review of these three stages.

2.1 Historical Background: An Unfortunate Precedent

The 1870s marked the inception of telecommunications in China. Due in large part to the undeveloped economy of China and its very weak military position at that time, foreign operators early and effectively dominated this important sector. In June 1871, the Danish Great Northern Telegraph Company (GNTC) constructed a 2,200-knot cable from Vladivostok in Russia via Nagasaki in Japan to Shanghai in China, which was the first ever telegraph circuit in China. According to an agreement between Denmark, Britain, and Russia, the GNTC and the British Eastern Extension Australia and China Telegraph Company (EEACT), whose shareholders mainly consisted of small groups of capitalists and royal family members of the above three countries, shared the rights of telecommunications provision in China from 1873 to 1899 [1]. According to the agreement, the coastal area north to Shanghai belonged to the GNTC, while the coastal area south to Hong Kong was under the control of the EEACT. Both companies jointly controlled the coastal area between Shanghai and Hong Kong. Thus armed, the GNTC and EEACT somewhat impertinently attempted to restrict the rights of the Chinese government to establish its own submarine telegraph networks. When China was preparing to build such telegraph networks in June 1881, GNTC presented a proposal to the then Qing Dynasty government claiming exclusive rights for providing telegraph services in the territories where it had installed submarine and land cables. Under the terms of this (approved) proposal, no other operators, including those from China, would be licensed to install cables in these areas in the ensuing 20 years [2]. A weak Chinese state was confronted by a strong company constellation.

The GNTC and EEACT became more aggressive after the joint invasion of China by eight Western countries (Austria, Britain, France, Germany, Italy, Japan, Russia, and the United Sates) in 1900. By taking advantage of the chaotic situation in China and the damage to the Chinese telegraph networks, the two companies expanded their business in the name of repairing the damaged circuits. The Chinese Directorate General of Telegraphs was forced to sign the contract to pay the bill for repairing the Chinese Yantai-Dagu-Shanghai submarine cable, which was twice as much as the actual cost for the repair service it did not order, and to relinquish all its rights to the two companies for 30 years before the bill was paid with 5% interest. Additionally, all of the previous contracts between the Chinese Qing Dynasty government and the two companies were automatically extended to the end of 1930 [2].

The GNTC and EEACT also proposed to the Qing Dynasty government and obtained approval to repair the damaged Tianjin-Beijing land cable and agreed to return it to China after a peace agreement was reached. The two companies, however, used this militarily important cable to bargain with the Qing Dynasty government after the war and obtained the right to lease two other cables, namely the Beijing-Qiaketu cable to GNTC and the Beijing-Dagu cable to the EEACT, before it returned the Tianjin-Beijing cable to China. In this way, China lost its control over international telecommunications as the Beijing-Qiaketu cable was the gateway connecting China to Europe via Russia. In this way, the two companies successfully extended their business to the land cable sector [2, 3].

The dominance of foreign operators set an unfortunate historical precedent—one, moreover, that led to a highly conservative telecommunications policy position on the part of later Chinese governments. They were to view telecommunications as of critical governmental and military importance. The most enduring policy ramification of this early period of foreign domination was that it encouraged extreme reluctance to allow subsequent market incursion on the part of foreign corporations, especially network operators. Early negative experiences with foreign companies thus taught the Chinese that such companies were exploitative and threatened the sovereignty of the Chinese telecommunications system [3].

As with the case of the telegraph, it was foreign operators that began to provide telephone service in China. The GNTC launched its telephone service in Shanghai on February 22, 1882, which was followed by a British company (Shanghai Mutual Telephone Association) 2 months later. In 1983, another British company, the China and Japan Telephone Company, bought the GNTC's network and consolidated it with that of the Mutual Telephone Association. The new company provided telephone service in Shanghai for 18 years until it failed in bidding against the Shanghai Mutual Telephone Company in 1900.

Also in 1900, a Danish businessman, Care H. O. Poulson, built China's first long-distance telephone line between Tianjin and Beijing. In 1904, the Beijing Directorate General of Telegraphs contacted Poulson with a view to handing over the telephone service to the Chinese government. After 50,000 silver yuan was paid, the transaction was made. On the basis of this service, the Beijing Telephone Directorate General was established, as was the Tianjin Telephone Directorate General. Public long-distance telephone service between Beijing and Tianjin began in 1905.

The importance of telecommunications was gradually realized by the Qing Dynasty government, and efforts were devoted to the development of telegraph and telephone services. The military telegraph circuit connecting Gaoxiong, Taipei, and Jilong in Taiwan built in 1877 was the first telegraph circuit that was built independently by China. In 1879, a telegraph linking the office of the head of the Northern Navy in Tianjin with the Dagu and Beitang fortresses and the Tianjin weaponry factory was installed. In December 1881, the telegraph circuit between Shanghai and Tianjin was constructed, becoming the first one to provide a public telegraph service. By 1890, local telephone service was available in Beijing, Tianjin, and Nanjing [2].

Because of insufficient investment funds, the Qing Dynasty government adopted a policy of commercial operation under governmental supervision at the beginning of the last century [2]. On November 6, 1901, the government approved the establishment of the Ministry of Posts and Transportation as an administrative regulator. From 1908, the ministry began to nationalize the telecommunications industry, and within a year all commercial operations had been bought.

In 1911, the Qing Dynasty was overthrown and China entered an era of political instability that lasted until 1949 when the Chinese Communist Party acceded to power. The period from 1911 to 1949 began with internecine strife between warlords followed by an 8-year invasion and occupation by Japan. This was succeeded by protracted civil wars between the Communist and Nationalist Parties.

In 1928, the Nationalist Party established a new government in Nanjing after defeating the Beiyang warlords. Telecommunications administration fell under the aegis of a Construction Commission and was later transferred to the Ministry of Transportation. Under pressure from the upsurge of anti-imperialism and nationalism, the government began to take actions aimed at restoring China's control over its telecommunications systems. Two radio stations were built and developed in Shenyang and Shanghai in 1928 and 1930, respectively, which together broke the foreign monopoly operation of international telecommunications services. In the meantime, the government began to negotiate with the GNTC and EEACT on reclaiming leased cables and terminating the 30-year submarine cable contracts. In 1933, the exclusive right of foreign submarine cables to terminate in China was revoked. In May 1934, China reclaimed all the leased cables [2, 3].

In 1937 Japan invaded China and occupied Chinese territory for 8 years. This was a destructive period for the development of telecommunications in China. For example, the total number of local telephone subscribers dropped from 55,683 in 1936 to 7,918 in 1944, while the total length of long-distance circuits dropped from 52,245 to 4,085 km in 1944 [2]. A 4-year civil war between the Communist and Nationalist parties immediately followed World War II, which allowed no time for the telecommunications sector to recover.

When the Communist Party won the war and founded the People's Republic of China in 1949, there were a mere 20,000 telephones and 2,800 trunks for long-distance telephone service throughout the entire country. The infrastructure was fragmented under the management of different interest groups. Systems lacked interoperability and compatibility, as equipment was made and supplied by different companies. The country lacked a nationwide backbone telecommunications network at that time [4, 5].

2.2   Telecommunications in a Centrally Planned Economy: 1949 to 1978

In November 1949, following the founding of the People's Republic of China, the Ministry of Posts and Telecommunications (MPT) was formally established. As a section of the Ministry, the Directorate General of Telecommunications (DGT) was set up on January 1, 1950, and placed in charge of almost all aspects of telecommunications, including administration, regulation, operations, human resource management, financial management, and equipment supply. In September 1950, the MPT was restructured along the orthodox bureaucratic lines of the socialist organizational model of the former Soviet Union. Each department within the MPT was responsible for the control of one of the following specific functions: financial management, human resource management, development strategy, regulations—both for posts and telecommunications. The DGT, as one of the departments, was placed in sole charge of coordinating operations of the telecommunications network.

The MPT acted as the headquarters of the national posts and telecommunications sectors and was mainly responsible for formulating development plans and coordinating nationwide networks. It was provincial Posts and Telecommunications Administrations (PTAs) that conducted routine network operations. The PTAs were subordinate branches of the MPT at the provincial level and were responsible for deploying all regulations and strategies of the MPT in each individual province. The PTA had corresponding sections subordinate to every department of the MPT. For example, each PTA had a financial section that was responsible for communicating and reporting to the Department of Finance of the MPT. In this way, a vertically integrated posts and telecommunications complex was set up nationwide in China, independent from provincial and local government. The organization appears to have exhibited classic bureaucratic dysfunctions, most notably upward referral of decision making and consequent overcentralization.

In June 1958, under the influence of the "great leap forward" strategy proposed by Mao Ze-dong, the State Council (China's Cabinet) agreed to a proposal emanating from the MPT to hand over all local posts and telecommunications enterprises to provincial and local governments with the exception of the telecommunications system in Beijing and the national backbone trunk network. This appears to have been consistent with one of the major defining political phenomena of the period: a ruthless attack on the "professionals who constituted middle administration in the national bureaucracy and middle management in industry" [6]. The process began in July of that year. In practice, chaos occurred, since in some places the enterprises were handed over to ill-equipped and inexperienced local governments in towns and villages, which led to a lack of coordination, poor management, and a general decline in telecommunications quality. As a result, the MPT had to make further adjustments in August and September 1959 and withdraw the power of development planning, capital budgets, equipment supply, and manufacturing from local and provincial government control. These powers reverted to ministerial and provincial levels under the centralized administration of the MPT.

By 1960, the short-lived great leap forward was rejected as a failure, which led to organizational dislocations that necessitated a further process of readjustment. As a result, a policy flip-flop took place in December 1962 when the Central Committee of the Chinese Communist Party (CCCCP) agreed with a report from the MPT arguing for full reversion to the preexisting centralized administrative system of posts and telecommunications. The revised system commenced operations in late 1962 and was similar in essentials to the administrative system prior to the great leap forward. Henceforth, powers of commencing and operating services, formulating rules and regulations, constructing infrastructure, defining tariff rates, managing finance, and supplying equipment were once again returned to the authority of the provincial PTAs and MPT.

This system lasted for a significant period until November 5, 1969, when the State Council agreed with a joint proposal by the MPT and the Ministry of Railways and Transportation [7]. According to this proposal, the MPT was dismantled starting December 1, 1969, the Directorate General of Posts (DGP) was joined together with the Ministry of Railways and Transportation, and, in turn, the DGT was amalgamated with the military under the administration of the General Strategy Bureau of the People's Liberation Army (PLA). This decision was beneficial for the postal service since its distribution logistic was similar to that of railways and other transportation systems. The unification significantly improved the synergy between the postal service and the transportation process. As a result, the postal service developed very rapidly [7].

This new scheme, however, was destined to last for only a few years. First, because telecommunications was still viewed as an instrument of governmental administration and national defense rather than as a commercial entity in its own right, its full development as a service industry had not been given high priority. The policy adopted in China had been to subsidize deficits in telecommunication services with profits from postal services by bundling together the administration of the postal services with the telecommunication services after 1949. The separation of posts and telecommunications (a process that began in 1969) left telecommunications in a very difficult financial situation [8]. Second, the PLA perceived the public telecommunications system as a burden rather than as a cash cow, and announced its intention of extricating itself from the telecommunications system.

For these reasons, the State Council and Central Military Committee jointly issued an "Announcement Concerning Adjusting the System of Posts and Telecommunications" in May 1973. Following this pronouncement, the MPT was reestablished on June 1, 1973. In the provinces, PTAs were also reestablished. Under the administration of provincial PTAs, posts and telecommunications were operated jointly in most cities and towns except in provincial capitals and in three cities that were placed directly under the jurisdiction of the central government (Beijing, Shanghai, and Tianjin), where they were operated separately due to the heavy demand for both postal and telecommunication services. The MPT and provincial governments jointly controlled provincial PTAs. Local posts and telecommunications enterprises were placed under the dual administration both of provincial PTAs and provincial governments. All revenues were handed over to the provincial governments. This structure existed until 1979 [8].

During the period of the centrally planned economy prior to 1978, the telecommunication industry had been treated as a monopoly subject to the bureaucratic authority of the state. This resulted from the assertion of economic, military, and national defense rationales, with military arguments taking a preeminent position. Entirely consistent with this orthodoxy, the MPT adopted a highly centralized form of administrative solution to the telecommunications problem. As a result, the telecommunications system was operated and managed on a nationwide basis in a semimilitary style. Administrative norms took precedence over a more market-oriented management system.

The consequence of such orthodoxy was a general stultification of telephone development. The adoption of a reactive administrative system rather than a proactive management structure discouraged incentivization within the organizations comprising the telecommunications system. Between 1949 and 1978, developments were largely restricted to backbone network expansion connecting major cities and sites of strategic importance. The luxury of the residential phone was only made available to senior government and military officials as telephone installation was viewed primarily as a symbol of political status rather than a commercial service based on the principle of universality of provision. As in other communist systems, elitist patterns of allocation ensured that priority was given to highly placed party officials and their families.

Under a centrally planned economic system, improving efficiency and making profit were definitely not the main objectives of the operating firms in the telecommunication industry. The MPT launched several income and expenditure management schemes, but none of them created much incentive for operating firms to improve performance levels. For example, in the early 1950s the organizations handed over their income to the MPT or the provincial PTA every 3 or 5 days and claimed back their expenditure retrospectively. In the late 1950s, the MPT implemented a system based on appropriation according to budget. Under this system, constituent organizations would propose their budget to their provincial PTA before the start of a new fiscal year. Authorized by the PTA, they would then be issued an appropriation quota memorandum (command) on a monthly or seasonal basis. Following receipt of the memorandum, they could keep part of their income while transferring the remainder to the provincial PTA. From 1969 to the early 1980s, the MPT implemented a difference management system (*difference* referring to the gap between income and expenditure). According to this system, firms would forward part of their income-expenditure difference annually according to a predecided quota, and any surpluses after this settlement would be shared between the provincial PTA and the firms.

Unfortunately, all the above schemes failed to motivate individual enterprises to highlight efficiency-improvement objectives as key components in their corporate strategy, although this was the clear intention of the difference management system. Under the scheme adopted in the early 1950s, which was labeled the treasury system, enterprises had no incentive to control their budget as they could always claim back their expenditure. They were effectively indifferent to concerns over income as they in no way directly benefited from it. Under the appropriation according to budget system, the fiscal situation of an enterprise depended on how much budget was allocated to it. In this case, as has been witnessed in other industrial sectors in socialist economies, vertical relationships between superiors and subordinates and those between planners and enterprise managers were critical and normally involved extensive bargaining [9–11].

The difference management system, while intended to motivate enterprises to improve efficiency, met with serious implementation problems due to the specific characteristics of telecommunications. Normally in telecommunications, more than two parties participate in the complete production process (e.g., in the case of long-distance telephone service). The user, however, will only pay the bill to the originating party, while the terminating party's contribution and cost will not be compensated. For example, for long-distance telephone services between Beijing and Tibet before the 1980s, most of the calls originated in Beijing; hence, in most cases only the telecommunications office in Beijing was paid rather than that in Tibet, although it was the telecommunications office in Tibet that terminated these calls. In this case, the difference between income and expenditure could not accurately reflect the contribution and efficiency level of a local telecommunications enterprise, and like scheme two, the issue of how to establish a favorable quota for difference remained a constant struggle against informal institutional constraints, such as personal relationships between managers and their supervising governmental officials.

Because of the above factors, enterprises in China's pre-1978 telecommunications industry lacked autonomy and growth potential. There were no incentives to pursue such objectives as efficiency and profit. The only objective of telecommunications organizations in this industry "tend[ed] to be in fulfilling the plan quota and thus winning recognition from administrative superiors" [12]. As a result of the above dysfunctional system, coupled with political interference emanating from the so-called great leap forward movement and the Cultural Revolution, the entire industry was forced to operate at suboptimal levels of efficiency. As a consequence, the telecommunications industry suffered serious losses for 9 years during the period from 1966 to 1978. The telephone penetration rate (the total number of telephone lines per 100 inhabitants) was only 0.38% in 1978, a rate that lagged behind more than 150 countries around the world [8, 13].

2.3 Telecommunications Reform from 1978 to 1994

From an institutional perspective, the inception and growth phases of organizations coming into existence and their evolution are fundamentally influenced by the institutional framework within which they are embedded [14]. The institutional framework can be defined as a set of fundamental political, social, and legal rules that establish the basis for production, exchange, and distribution [15]. Such an institutional framework affects the actions of organizations by constraining the organizations within a set of cultural expectations. Thus, organizational activity is either deemed legitimate or unacceptable to the extent that it accords with patterns of expectations set by the parameters of the institutional framework [16]. For example, Davis, Diekmann, and Tinsley [17] have suggested that institutional factors in specific cases brought about the decreased use of the conglomerate form of organization in the 1980s because the very idea of this form was no longer deemed legitimate.

A characteristic feature of the institutional framework of China's planned economy has been the normalization of central planning and bureaucratic control [18]. As noted earlier, such central planning and control have constrained the behavior of telecommunications enterprises, leading to low efficiency. Additionally, the political objective of the government in using telecommunications as an instrument of both the government and military clearly inhibited the utilization of commercial incentives within the industry.

Growing awareness of the necessity of hastening the development of telecommunications led to the realization that improvements were essential in the centrally planned economic system. The turning point was in the late 1970s when China embarked on a major program of economic reform (beginning in 1978). The government initiated an ambitious strategy aimed at the realization of four sectoral modernizations: industry, agriculture, defense, and science. Central to such ambitions was the recognition that telecommunications was a key facilitator of industrial and commercial progress. The government clearly realized that a poorly developed telecommunications infrastructure acted as a major demotivator in attracting investment into local economies. Low penetration rates coupled with the fact that only a limited number of cities could provide international telephone service inhibited investment by multinationals [19]. In addition, with its open door policy, China belatedly realized that the world had entered the information age and that telecommunications should no longer be seen purely as an instrument for governmental and military use, but a kind of commodity—not something unproductive, but a profitable area in its own right and a critical component in a drive to promote the development of other industries as well.

Such arguments were fully articulated at the Seventeenth Annual General Meeting of the MPT held in March 1979. At this keynote meeting, the MPT announced its full support for the rapid development of telecommunications services and infrastructure and acknowledged the salience of telecommunications for the modernization drive. This report acknowledged, for the first time, that telecommunications was not simply another branch of government but an industry in its own right. Even more radically, the report insisted that the telecommunications provider should embrace an entrepreneurial, rather than a bureaucratic, culture and that telecommunications enterprises should not only fulfill the government plan but also meet the demands of the market [20]. All revenues would be centrally managed by the MPT, and the system of posts and telecommunications should reside under the unified MPT administration. Provincial governments were no longer to be directly responsible for the development of posts and telecommunications, and routine management of telecommunications should, in consequence, be transferred to the MPT's provincial branches (the PTAs). The influence of the provincial government would henceforth be very weak [20].

It proved easier to stipulate these goals than to achieve them. To speed up the development of telecommunications, the most difficult problem lay in capital formation. Investment in the telecommunications infrastructure by the state government via the MPT was limited due to the state government's weak fiscal power. Provincial PTAs had to raise funds themselves to support their development. In such a situation, the power of decision making and financial responsibility had to be partly delegated to PTAs by the MPT.

In order to initiate a radical process of decentralization, the State Council issued a directive (No. 165) in 1979. This directive for the first time stipulated that the functions of governmental administration and business operation should be separated in the posts and telecommunications industry. It also suggested that the postal and telecommunication services should be separately administered and each enterprise should have its own accounts and be financially independent. This innovative directive proved critical to the development of Chinese telecommunications [20]. The postal and telecommunication services, however, were not separated until 1998. Since telecommunications was becoming increasingly prosperous, while the postal services remained relatively backward, it was strongly contended that the postal services could be cross-subsidized by telecommunication services if they were administrated in the same ministry [7].

To transform telecommunications operation from political instrument to commercial business, previous approaches to enterprise management were to be the subject of reform. As Boisot [21] has noted, the lack of clarity in specifying contractual rights and responsibilities between administrative institutions and enterprises constitutes a fundamental problem in socialist economic systems. Recognition of this problem in the early 1980s led the Chinese government to implement key reforms in the telecommunications industry including the decentralization of administrative power to lower government echelons, the delegation of responsibility for performance to enterprise management, and the adoption of a variety of incentive systems [22]. Key reform schemes in the telecommunications sector (which have proven to be very successful since 1978) are reviewed below.

2.3.1 The Contractual Responsibility System

The so-called Contractual Responsibility System evolved from experience gained in China's agriculture reforms, which began in 1979. Success achieved in agriculture encouraged the Chinese government to transplant this system into industry, but this was difficult given the relative complexity of the industrial system.

The government's intention was to clarify responsibility for success and failure at all levels of the industrial hierarchy down to factory workers, and then decentralize the power to these levels accordingly. It commenced with a contract system, allowing direct interenterprise transactions and giving enterprises control over part of the surplus output above the contracted level. This marked the real commencement of decentralization of government power to enterprises, although, in effect, the government retained a significant component of its authority through a planned economy structure. By 1982, the contractual responsibility system had been applied to more than 300 enterprises.

In 1984, the State Council decided to further extend the decision-making freedom of state-owned enterprises in 10 areas comprising production planning, product marketing, pricing of products outside the state plan, material purchasing, use of surplus funds, disposal of assets, organization, personnel management, payment systems, and transactions between enterprises. In the same year, the CCCCP published a document entitled "China's Economic Institutional Structure Reform." It specified that enterprises should become "independent and responsible for their own profit and loss and capable of transforming and developing themselves ... [acting as] legal persons with certain rights and duties." Their directors would have conferred upon them full responsibility for their performance under a unified management system. These responsibilities included the following:

- Determining the enterprise's operation and organization, including more flexible and diversified forms;

- Planning its production, supply, and marketing in tune with market mechanisms;

- Setting prices within limits prescribed by the state;

- Retaining certain budget funds;

- Stipulating recruitment, personnel, and reward policies.

In this so-called Director Responsibility System, enterprise directors have enjoyed higher autonomy than previously. Directorial incentives and motivation have been sharpened. Promotion and succession have been linked to performance and achievement. Simultaneously, directors have been subjected to stronger pressures, both from the authorities supervising their performance and from employees in the enterprise heavily dependent on their commitment (the fate of the employee, in turn, being more closely allied to that of the enterprise). These incentives were intended to motivate directors to achieve higher production standards, while transforming erstwhile party or government agents into entrepreneurs, who can run their enterprises independently according to modern management techniques. Training programs organized by the central government bolstered this philosophy. As a result, business efficiency was greatly enhanced.

In order to formalize the Director Responsibility System, a Contractual Management Responsibility System was also established in 1986 in many state-owned enterprises. Although it was applied in different industries, several common features were displayed. They included the following:

- The governance of relations between enterprise decision-makers and supervisory agency by some form of explicit contract;

- The participation of an enterprise director in a contract on behalf of an enterprise while assuming personal responsibility for the attainment of enterprise goals;

- Some assumption of risk by the contracted parties, along with variability in their rewards linked to the performance of the enterprise;

- Contract terms extending over several years [23].

The MPT was an early starter in the deployment of the Contractual Responsibility System. From 1980, all the provincial PTAs signed contracts with the MPT according to the principle of "twin commitments and single linkage." The first commitment means the enterprise is obliged to hand over an agreed amount of annual profit and tax (although profits were not accurately estimated—a problem to be highlighted later on in this chapter) while retaining a proportion of any surplus it achieved above the contracted level. The second commitment ensures that the enterprise must guarantee to invest in order to increase asset values and develop technology by an agreed amount during the period of the contract. The single linkage means that up to 70%

of the increase in overall profit can be used to fund salary increases, linking individual staff rewards with performance.

In 1982, the MPT adopted a Three-Year-Term Profit Rolling Contract System. This system defined an annual growth rate of overall profit for 3 years, each enterprise's performance being assessed according to the contract every year. According to the result of assessment, terms of the contract would be revised and extended for another year. In this case, the enterprises were always rendered with the profit objectives for 3 consecutive years. This overcame the disadvantages of short-term contracts since some directors subject to such contracts typically adopted a short-term approach and management clearly lacked long-term development strategies.

Since early 1986, the MPT, along with other industries in China, deployed the Contractual Management Responsibility System. Under this, directors of the provincial PTAs signed contracts with the MPT every year. After negotiation, such objectives as traffic, revenue, service quality, and efficiency targets were contractually defined, which led to quantifiable reward and penalty measures for the enterprise. In order to prevent short-term thinking and to provide incentives to directors, 4-year appointments were also signed. When the appointment period expires, the performance of the director is reviewed. Based upon the result of this review, the authorities decide if the director should be reappointed, promoted, or dismissed.

In October 1986, the contracting principle was extended to the internal management of enterprises ranging from employment to production. It abolished life-tenure or the "iron rice bowl" for newly recruited employees and made any contract renewal conditional upon an individual employee's performance and the needs of the enterprise. At the same time, every employee signed a contract with the enterprise specifying the required quality and quantity of a product, and the rewards/penalties for performance/nonperformance. This laid stress upon worker responsibilities in the production process alongside corresponding worker power and rights. By this means, it was intended that every employee should be clear about his or her responsibilities, rights, and interests. Thus, a new incentive structure linking rewards more directly to output was established.

The system of responsibility has effectively improved the efficiency of telecommunications operation following its adoption in the early 1980s. Figure 2.1 indicates that labor productivity has increased consistently since the reform commenced in the early 1980s. This trend has continued. The MPT, for example, enjoyed a 37% increase in overall labor productivity in 1992 compared with 1991 [24], and a 51.9% increase in 1993 compared with 1992 [25].

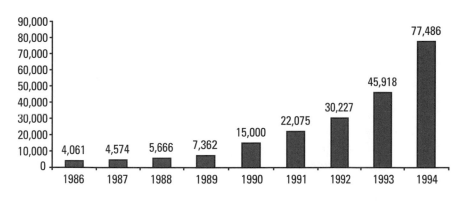

Figure 2.1 Labor productivity of the MPT (*unit:* yuan/person). (*Source:* MPT.)

2.3.2   The Economic Accounting System

The Contractual Responsibility System has proven to be a very successful management scheme. There still exists, however, a problem of management accountability—that is, how to quantify operational performance clearly and accurately. If this issue is ill-defined, then the contract cannot be accurately specified, and the directors' performance cannot be fairly assessed, effectively rendering the contract worthless.

In other industries, a general economic accounting system is applicable. For telecommunications, however, a special accounting system is required due to its unique production characteristics.

The most evident and discrete characteristic of telecommunications is that the whole process of telecommunication, especially long-distance telecommunication, is a relay process, conducted across the network, requiring the joint contributions of at least two parties. A complete long-distance telecommunications process is described in Figure 2.2.

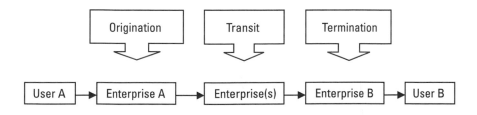

Figure 2.2 Long-distance telecommunications process.

User A accesses the local telecommunications carrier (enterprise A) and starts the process of communication with user B. Enterprise A originates the call and sets up the connection with enterprise B, whereas enterprise B finalizes the connection process by terminating the call at user B's end. Unless the two enterprises are directly linked, messages are transmitted from enterprise A to enterprise B via a series of transit enterprises.

It is obvious that each carrier enterprise provides a critical link in this process because communications are blocked if any one link is out of order. It is therefore impossible to complete a communication without cooperation between these telecommunication enterprises. The customer, however, is only charged by the origination carrier, while the transit and termination carriers are not directly paid by customers. How to clarify the costs in this process and how to reallocate the total revenues among all participating enterprises are major problems featuring in any telecommunications accounting methodology. Otherwise, any incentive for the transit and termination carriers is frustrated by their contributions remaining uncompensated or inappropriately compensated.

The difference management system mentioned earlier, which was implemented between November 1969 and the early 1980s, was intended to place some pressure on the enterprise and its directors to focus on operational performance in order to achieve the agreed-upon difference. It also aimed to stimulate their incentives, as their enterprise could benefit from the attainment of a surplus from the difference. These intentions, however, were not fully satisfied for the following reasons:

1. As the above-mentioned example of long-distance telephone services between Beijing and Tibet demonstrated, a key characteristic of long-distance telecommunication service is that it is a relay process conducted along the network, but where the original customer only pays the initial carrier. Carriers involved in transit and termination are not directly paid. In this case, the revenues of those carriers do not reflect their performance or contribution.

2. There is a danger in encouraging the enterprise to focus on call-origination only, as the transit and termination services are not compensated. This may lead to a lack of coordination within the network, to the detriment of overall telecommunications quality and revenue.

3. Even for identical services, the cost and expense of operations vary according to the geographic location of the enterprise. For example,

places with a poorly functioning economy may have only limited demands for a full-fledged telecommunications service. Lack of economies of scale here result in high averaged operational costs, a factor that this system does not take fully into account.

4. For telecommunications, the difference between income and expenditure within one single enterprise does not actually equate to profit. Only the overall income-expenditure difference within the nationwide network is equal to overall profit. If profit cannot be calculated in the enterprise, then corresponding business and operational indicators, such as labor efficiency and ratio of profit to production output, cannot be properly measured; hence, operational results cannot be properly benchmarked.

For the above reasons, the difference management system, in fact, constituted a bureaucratic system with many disadvantages. Operational performance could not be properly measured. Moreover, this mechanism, by encouraging call origination only, failed to facilitate cooperation among carriers in different locations.

Realizing these disadvantages, the MPT embarked upon the implementation of an economic accounting system. This system was tested in Hubei and two other provinces during 1983 and 1984. Very encouraging results were achieved, and the system was adopted nationwide in 1985.

The major breakthrough at the heart of this system lay in defining how to calculate an individual provider's own income, the so-called enterprise-owned revenue (EOR). This was actually a method of reallocating the whole network's income, especially long-distance service revenue, according to an individual enterprise's contribution towards the network operation. If a local telephone provider's assigned revenue could be calculated accurately, then the profit or loss could also be measured within relevant parameters. Based upon this calculation, the directors' performance could be assessed more accurately, and responsibility, reward, and penalty could be more clearly defined in their contracts. Only by using this methodology could the Contractual Responsibility System become a practical and feasible management scheme for telecommunications.

In order to calculate the EOR, the MPT classified the nature of the call transferred through the originating provider, transit enterprise, and call terminator as export product, transit product, and import product, respectively (Figure 2.3). For each product there is a fixed price, which is determined by cost allocations conducted by the MPT. The price is fixed for about 5 years

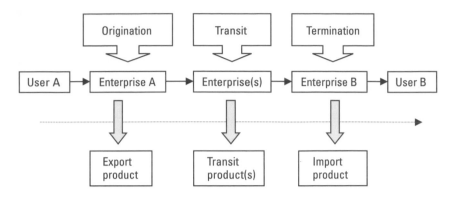

Figure 2.3 Classifications of telecommunications products.

and is then reviewed by the MPT. Table 2.1 shows part of the price catalog released in 1991 [26]. All the long-distance traffic in telecommunications, including export, transit, and import, is recorded by the enterprise. Then the sum of the traffic of each product times its corresponding price equals the total long-distance telecommunication service revenue of that enterprise. By using this method, the contribution of each enterprise to each procedure can be calculated and charged. Enterprises increased their attention to transit and import services since they obtained revenues from them as well. In this way, cooperation among individual enterprises is encouraged, and both the volume of traffic and quality of service has improved greatly. Income from local network services is retained by the local enterprise and is not reallocated.

Since financial capital is critical to the further development of advanced telecommunications, any growth in income is important. The MPT, therefore, encouraged local enterprises to increase their income by allowing them to retain a proportion (currently 40%) of their increased income compared with previous years.

The total EOR of an enterprise can be calculated as follows:

Total EOR = export products traffic * export products prices + transit products traffic * transit products prices + import products traffic * import products prices + all income from local network services + (income of this year − income of last year) * retaining rate (40%)

It can thus be seen that this method has the following advantages:

Table 2.1
Economic Accounting Prices for Telecommunications (1991)

Name of Product	Accounting Unit	Unit Price	Name of Product	Accounting Unit	Unit Price
Export telegraph	Number	0.5607	Export long-distance telephone	Number	1.7345
Import telegraph	Number	0.5455	Import long-distance telephone	Number	2.1188
Transit telegraph	Number	0.2267	Transit long-distance telephone	Number	2.4614
Export facsimile	Number	6.8648	Long-distance cable	Km	15041.38
Import facsimile	Number	89.7907	Satellite television channel	Hour	538.366
Export Telex	Number	1.3254	Satellite telephone channel	Circuit/day	45.2288
Import Telex	Number	0.8423	Satellite telegraph channel	Circuit/day	4.7689
Transit Telex	Number	0.4054	Export teleconference	Per time	14.8075
Radio transmission	kW hour	10.1133	Import teleconference	Per time	27.1951
Radio receiving	kW hour	12.4663	Transit teleconference	Per time	27.4445

- For the first time it offers a way of measuring the operational performance of each entity, allowing profit or loss to be calculated. It provides a quantitative method for management and thus makes the Contractual Responsibility System feasible in the telecommunications sector.
- By remunerating them for their transit and terminating contributions, it encourages individual enterprises to cooperate in the process of telecommunications, which improves network cooperation and simultaneously improves service quality, leading to an increase in traffic over the whole network.
- It provides an incentive for telecom enterprises to increase their income, which facilitates recovery of previous large investments and allows the collection of sufficient finance for further investment.

- It creates incentives for the effective self-development of enterprises, as they can benefit both from their contribution to network traffic and their share of increased income.

As mentioned earlier, the geographic location of telecommunications providers impacts upon the cost of their operation. In order to solve this problem and redistribute income in a fair and equitable way, the 1985 scheme divided China's 30 provinces into nine categories according to their local objective conditions. Each category has a set of corresponding accounting prices. For provinces in adverse situations, the price for accounting purposes is correspondingly higher, which means they are compensated appropriately for their service compared with provinces in other more favorable areas. Unsurprisingly, this has created many disputes among these provinces as to who should belong to a category with higher accounting prices as opposed to allocation to a category with lower prices. At the same time, calculating nine categories of accounting prices was also complicated because of the considerable work involved in analyzing the cost allocations (the precise allocation of cost is a generic problem in all telecommunications systems).

In 1988, the MPT again readjusted its economic accounting system. Similar to that proposed in 1985, it comprised just one set of prices instead of nine, which simplified the problem of setting a price. In order to reallocate the revenue in a fair and equal way, the MPT proposed the so-called cost variance coefficient. Seven factors were chosen from more than 20 variables as the most relevant in affecting the cost of telecommunications services in an area [27]. The parameters helped to objectively describe the different conditions in each area. This was achieved through multifactor regression using seven factors including the ratio of mountain areas, the number of days without frost within a year (which is relevant to the issue of heating expenses), the number of residents in the city and town as a percentage of all residents in the whole region, the average consumption of telecommunications per square kilometer, the average GDP per capita, the density of railways and roads, and the average wage per employee. After the regression, each area was allocated a cost variance coefficient (CVC). For instance, the CVC for Beijing was 0.69 in 1988 while the CVC for Tibet was as high as 2.13. When the CVC is used to multiply the standard price, a special scale of prices for an individual area can be calculated. Then, the original formula of 1985 can still be used to calculate the EOR.

In fact, the Economic Accounting System played an important role in cross-subsidizing the high-cost area, even though providing universal service was not accorded a high priority on the government's agenda at that time. It may be contentious to argue that this is a kind of cross-subsidization, as the general principle of designing this scheme (in the view of one of the authors of this book, who participated in the formulation of this scheme in 1988), was to find the cost of each party's contribution and then to compensate it appropriately at reasonable prices. This simple principle, however, has, in fact, played a role in facilitating universal services. As operations in high-cost areas were fully compensated, operators could keep business going and afford to provide long-distance service to users at the same price as their counterparts in low-cost areas.

Since the application of the Economic Accounting System in 1985, the management of telecommunications has improved dramatically. According to Liu Wang-jin, Deputy Director of the Department of Finance in the MPT, the Economic Accounting System makes "the directors and employees feel that they are owners of the enterprise, and they begin to draw attention to the performance of the business operation. It has effectively enhanced the management's consciousness in improving management efficiency and service quality. Production costs have been effectively controlled and the net economic benefit accruing to the enterprise has greatly increased" [26].

2.3.3 Material Incentives

Before commencement of the reforms in China, people's expectations were linked to the so-called iron rice bowl, which effectively meant that everybody was guaranteed a job for life. The most obvious disadvantage of this system was that high productivity was not encouraged, as individual payment was not linked with performance. The government typically motivated staff by moral encouragement, such as organizing socialist work competitions and nominating model workers. After the Cultural Revolution, however, there was a crisis of belief among the Chinese people. Management schemes with political overtones were no longer effective, and money became the first priority for many. In this situation, material incentives were used to motivate employees, mainly in the form of bonuses. "In practice, a relatively egalitarian distribution of bonuses and other incentive payments continues to prevail, especially in the state sector. More progress appears to have been made towards linking payment to a person's level of responsibility, education and training, and to the profit performance of the enterprises" [9]. The amount of bonus that the enterprise was allowed to distribute and that to which an individual was entitled were clearly specified in the contract.

From 1985, the MPT began a process of salary reform. The purpose was to link the employees' incomes to the performance of their employing (telecommunications) enterprise, and hence provide them with an incentive for offering better customer service. In order to enhance cooperation throughout the network, the growth rate of salaries was initially only linked with traffic growth rates. The salary growth rate was determined by a formula based on 50% of the provincial traffic growth rate and 50% of the national traffic growth rate. This system was functional for promoting cooperation, but incentives to increase revenue (eagerly desired by the MPT for its further investment) were still not being effectively stimulated. In order to solve this problem, the MPT altered the formula to include revenue as an additional factor in 1993. Salary growth rate was then determined by a formula using 40% of the national traffic growth rate, 30% of the regional traffic growth rate, and 30% of the regional revenue growth rate. This formula not only encouraged enterprises to cooperate with operators in other regions for long-distance services by considering growth of all kinds of traffic (origination, transit, and termination), it also encouraged them to increase their own revenue as they could benefit from this growth as well. Experience has shown that this system works well, and employee income has increased steadily.

2.3.4 Flexible Financing Policy

Blackman [28] argues that it may be true that the considerable cost of telecommunications networks cannot be financed solely from most countries' public sector funds. This does not necessarily mean, however, that private investment has a major role to play in all countries. The case in China is a good example: As a developing country with GDP per capita below the world's average level, falling into the World Bank's low-income economy category in the early 1980s, it was clearly impossible for the infant private sector to itself finance the entire cost of an advanced telecommunications system. Additionally, privatization was an ideologically sensitive issue in the early 1980s when China had just started its transition from a centrally planned to a socialist market economy. The central government, however, also found it impossible to satisfy the financial needs of a capital-intensive telecommunications system that was hungry for investment.

Acknowledging this situation, the government issued an edict to the MPT in 1984 that encouraged it to adopt a policy of diversifying financial resources and exploring investment from central government, local government, corporations, and a proportion of private sector funds. In addition to this, the MPT has benefited from a variety of concessions from central government. This includes the policy of the "three 90 percents":

1. 90% of profit is retained by the telecommunications enterprises (in other words, the tax rate is 10%, much lower than the 55% tax rate for other industries).

2. 90% of foreign exchange (hard currency) earnings are retained by the enterprise.

3. 90% of the central government's investment is treated as nonre-payable loans [29].

The local telecommunications enterprises are also allowed to collect installation fees from customers applying for subscription to the public telephone network. The fee level is subject to the degree of demand. According to the Telecommunication Indicators Database of the International Telecommunication Union, China's installation charge was amongst the highest in the world. In Beijing, for example, the installation fee was 5,000 yuan (about $606). The installation fee can be regarded as pure income. Local revenue and investment were therefore heavily dependent on the installation fee. In Shanghai, for instance, nearly a quarter of the capital for network investment came from installation fees during the period 1980 to 1991 [30].

In addition, loans and other forms of finance provided by foreign countries have also been encouraged, including financial capital from Japan, France, Belgium, Spain, and elsewhere, mainly in the form of governmental soft loans incurring low interest. In 1994, for example, about 15% of total investment by the MPT was accounted for by foreign finance.

As a result of the above preferential policies, the financial resource pool was greatly expanded. Figure 2.4 shows the structure of MPTs financial resources in the 1993–1994 fiscal year (unattributed interview with an MPT officer, October 18, 1994).

The above four innovative management reform schemes have played a crucial role in improving telecommunications enterprise management. Labor productivity, revenue growth rate, average income per capita, and other indicators have demonstrated that these schemes are effective. Since the reforms were initiated in the late 1970s, MPT is said to have successfully overseen the most rapidly developing sector in China [31]. The country experienced an average annual growth rate of 7.83% in the number of telephone subscribers during the period from 1980 to 1985, rising to 17% for the period from 1986 to 1990. In 1991, this number grew to 23% over the previous year—a continuing exponential trend revealed in a growth rate of 36% for 1992, 52.8% for 1993, and 71.1% for 1994. Statistics show that the growth rate of telecommunication services in China is higher than the growth rate of the

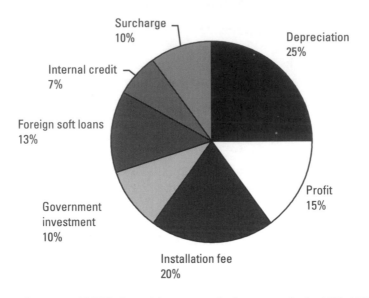

Figure 2.4 Structure of MPT's financial resources for investment in the 1993–1994 fiscal year.

GDP. The average annual rate of growth for telecommunication services was more than twice the GDP during the period from 1985 to 1990 [32].

It is therefore not an exaggeration to argue that these schemes (essentially proxies for privatization) have, at least to some extent, achieved similar results to privatization in improving internal enterprise management and financing necessary investment. Table 2.2 synthesizes the effects of these reform schemes.

As noted by Peng in his studies of other industries in China [12], these reforms "represent a major organizational innovation before fully-fledged privatization takes place." The contractual systems effectively conferred greater autonomy on enterprises from government. From a resource-based view [33], these telecommunication firms were then able and motivated to acquire and utilize resources with respect to production, technology, finance, and organization. Because of the rigid and vertically integrated organizational structure of the monopoly, however, the enterprises tended to exploit more resources for production and employee welfare rather than profit. This is consistent with the agency theory assertion that managers seek to maximize a utility function that contains status, power, security, and income as its central elements [34, 35]. Maximizing such a utility function may create a preference for inefficient empire building at the expense of profit maximization.

Table 2.2
Effectiveness of Telecommunications Reform Schemes in China

	Decentralization	Market Origination	Delegation of Responsibility	Creation of Incentives
Contractual Responsibility System	*	*	*	*
Economic Accounting System	*	*		*
Material Incentives			*	*
Flexible Financing Policy	*	*		

This indicates continuing residual problems with the state system in this transition phase.

In addition to searching for more productive investment, the implementation of the Economic Accounting System has motivated individual enterprises to maximize their production output. By using this method, the contributions and costs earned and incurred by each enterprise can, in theory, be calculated and charged. Arguably, enterprises have started to pay attention on transit and import services since they have obtained income from them. In this way, all provincial enterprises have been motivated to adopt a production maximizing approach by facilitating growth in the volume of traffic.

While neoclassical theory has characterized firms as profit maximizers, Granick [36] and Kornai [37] have forcefully indicated that a socialist state-owned-enterprise (SOE) is typically anything but a profit maximizer. The experience of the Chinese telecommunication industry between 1978 and 1994 provides further proof in support of this finding. Although individual firms had aggressively exploited resources for production and employee welfare, profit was paid scant attention. This situation was effectively underwritten by the norms and values of a closed administrative/bureaucratic system. Weaned on the verities of socialist production, such a system became increasingly incongruent with a world of changing technology, competition, and globalization.

Indications of the internally focused pathology at the center of this system can be seen in its rationale of prioritization. Rather than being concentrated on external marketing issues, significant amounts of resources were being used for the development of noncore business activities, such as establishing training centers, building staff apartments, and giving generous bonuses to employees. In this way, not only could the telecommunications industry continue to enjoy preferential support from the central government due to its artificially low profit rates, but also it ensured that it retained overall control of government imbued resources while engaging in a retreat from a situation of detailed central government financial control.

One of the grotesque features of such a process may be discerned from a research project conducted by the MPT in the early 1990s. This actually attempted to explore how the costs of the telecommunications operation could be inflated [sic] to leverage more support from the government. Such cost inflation was, in a Machiavellian sense, clearly designed to enable the ministry to better bargain with the central government for preferential policy by demonstrating that the MPT was still a low-profit or loss-making ministry (interview with a former research fellow of the MPT on July 8, 2001). This is a cynical example from the Chinese context of the kind of bureaucratic maximizing behavior that has been documented in several contemporary accounts of the bureaucratic phenomenon [38].

Another reason for creating noncore business sectors is to accommodate surplus employees. Because of technological advancement, many manual jobs have been replaced by computerized systems. A vast number of toll operators, telegram typists, and other less skilled employees have been withdrawn from the frontline. According to China's employment policy, they are permanent employees and should be looked after by the constituent organizations in the industry. Although the MPT has adopted a zero-growth personnel policy since the early 1990s, which means the total number of employees in each individual firm has been frozen, an increasing number of employees require repositioning. As a result, telecommunication firms have been forced to create many noncore business subsidiaries. These affiliated agencies have been treated particularly favorably. For example, for every newly recruited subscriber, the affiliated agency can receive up to 350 yuan in commission, while the nonaffiliated can receive up to 300 yuan. Profit that is generated by affiliated agencies is normally disbursed as employee bonuses across the entire organization or used for other informal business purposes. Income from these noncore business sectors is sometimes referred to as the organization's minitreasury.

From the late 1980s to the mid 1990s, maximizing enterprise self-controllable resources and accommodating surplus employees were twin objectives of corporate strategy in this particular phase of telecommunications history. These self-regarding inner-directed policies were the unintended consequences of the government's commitment to telecommunications development, the implementation of reform schemes, and the life-long employment policy. By the end of 1999, there were 3,545 noncore business utilities inside the incumbent, comprising 183,000 employees and a total volume of assets worth 36.17 billion yuan. These strategies obviously differed from good practice in capitalist economies where profit-maximization (or at least optimization) is the characteristic core of operations. Given this background against which major efficiency improvements were needed in the telecommunications industry, more radical reform schemes were eventually deemed necessary, with deregulation and liberalization high on the reform agenda.

2.4 Conclusion

The period leading up to 1994, as with the entire twentieth-century development of Chinese infrastructure, was one that was clearly eventful and punctuated by (eventually inexorable) pressures for change. This period, leading to the high water mark of 1994, is probably best characterized as one in which the increasing importance of telecommunications as a medium of commerce and means of information exchange became self-evident to the Chinese political elite. It is tempting to suggest that developments within the postwar telecommunications sector were corseted by their location within a tight system of bureaucratic constraints. The logic of administrative action in this sector owed more to Soviet than Western provenance, with the telecommunications sector dominated by hierarchical organizations focused more on internal procedure and practice than the demands of the marketplace. We suggested in Chapter 1 that the history of Chinese telecommunications in this sector reveals a pattern of operation that we have chosen to characterize as one of administrative governance—procedurally oriented and risk averse. As pressures for change in this locomotive sector began to grow (exponentially) throughout the world, and as China began to renounce some of the most obvious features of autarky—deep suspicion of outsiders, a fortress mentality, and the search for national self-sufficiency and self-reliance in security—the country's leaders woke to the challenge of change and began to construct the renunciation of previously held and sanctified beliefs. In consequence, the first traces of managerialism and the emergence of a system of market governance, stressing the virtues of competition, risk taking, and consumer demand, began to make their appearance in the telecommunications sector.

The fruits of this process—halting steps towards reform—resulted in some of the system changes documented in this chapter. Proxies for competition and privatization began to appear. This chapter has focused on some of the ways in which new systems of values and operational measures of performance laid the foundations for further change and development. It would be erroneous to argue that this process has been achieved unproblematically. Old ways of thought die hard, and the dependency culture fostered by the iron rice bowl attitudes is unlikely to disappear overnight.

What this chapter has attempted to demonstrate is that pressures for change have mounted over the past two decades in this and other key sectors within the Chinese economy. For all its self-publicity as a revolutionary society, China continues to display deeply conservative features. Its very history as a country of social engineering and reengineering summoned up in the great leap forward, for example, has accentuated the conservationist side of Chinese policy. The phrase *touching stones to cross the river* nicely encapsulates this side of Chinese policy thinking with its emphasis on limited experimentation and caution. This is strongly reminiscent of Charles Lindblom's [39] brilliant characterization of the contemporary policy process in many societies as one less of radical change urges rather than one in which new policy marginally succeeds its predecessor in terms of a process of "successive limited comparison" ensuring continuity and stability.

China's embrace of Western market ways of thinking in telecommunications and elsewhere is likely to be cautious and careful. To those critics eager for the country to shake off its bad old Socialist habits, the process of change may be viewed as too little, too late. For the discerning Sinologist, however, wonderment may lie in the distance that the country has traveled in the past two decades and the comparative advantage ironically conferred upon it by its second-mover status. To paraphrase Dr. Johnson, in terms of China's incorporation of the market mechanism and all that that entails, the wonder is "not that it isn't done well, but that it is done at all." The following chapters will analyze events after 1994 and will suggest that policy lessons from abroad have been well applied within the Chinese telecommunications policy community and subtly adapted to suit local Chinese circumstances.

References

[1]   Baark, E., *Lighting Wires: The Telegraph and China's Technological Modernization*, Westport, CT: Greenwood Press, 1997, pp. 186–189.

[2]   He, Z., "A History of Telecommunications in China: Development and Policy Implications," in *Telecommunications and Development in China*, P. Lee (ed.), Cresskill, NJ: Hampton Press, Inc., 1997, pp. 55–88.

[3] Tong, X. Y., *The History of Posts and Telecommunications*, Beijing, China: China Book Bureau, 1988.

[4] Ye, P. D., and R. Skillen, "Telecommunications in China," *IEEE Communications Magazine*, Vol. 31, No. 7, 1993, pp. 14–16.

[5] Wan, S. D., "An Overview of Telecommunications in China," *IEEE Communications Magazine*, Vol. 31, No. 7, 1993, pp. 18–19.

[6] Schurmann, F., *Ideology and Organization in Communist China*, Los Angeles, CA: University of California Press, 1968.

[7] Wu, X. N., and H. R. Yang, *The Organization and Administration of Postal Service*, Beijing, China: People's Posts and Telecommunications Press, 1989.

[8] Liang, X. J., and R. Z. Yang, *The Organization and Administration of Telecommunications*, Beijing, China: People's Posts and Telecommunications Press, 1987.

[9] Child, J., *Management in China: During the Age of Reform*, Cambridge, UK: Cambridge University Press, 1994.

[10] Lu, Y., *Management Decision Making in Chinese Enterprises*, London, England: Macmillan, 1996.

[11] Rona-Tas, A., "The First Shall Be Last? Entrepreneurship and Communist Cadres in the Transition from Socialism," *American J. Sociology*, Vol. 100, 1994, pp. 40–69.

[12] Peng, M. W., *Business Strategies in Transition Economies*, Thousand Oaks, CA: Sega Publications, 2000.

[13] Zhu, Y. N., and G. L. Li, *Economics of Posts and Telecommunications*, Beijing, China: Beijing University of Posts and Telecommunications Press, 1990.

[14] North, D. C., *Institutions, Institutional Change and Economic Performance*, Cambridge, MA: Harvard University Press, 1990.

[15] Davis, L. E., and D. C. North, *Institutional Change and American Economic Growth*, Cambridge, UK: Cambridge University Press, 1971.

[16] Aldrich, H., and C. M. Fiol, "Fools Rush In? The Institutional Context of Industry Creation," *Academy of Management Review*, Vol. 19, pp. 645–670.

[17] Davis, G. F., K. Diekmann, and C. H. Tinsley, "The Decline and Fall of the Conglomerate Firm in the 1980s: The Deinstitutionalization of an Organizational Form," *American Sociological Review*, Vol. 59, 1994, pp. 547–570.

[18] Peng, M. W., and P. S. Heath, "The Growth of the Firm in Planned Economies in Transition: Institutions, Organizations, and Strategic Choice," *Academy of Management J.*, Vol. 21, 1996, pp. 492–528.

[19] Ure, J., "Telecommunications, with Chinese Characteristics," *Telecommunications Policy*, Vol. 18, No. 3, 1994, pp. 182–94.

[20] Yang, H. R., and X. N. Wu, *The Postal Service in The New China*, Beijing, China: Beijing University of Posts and Telecommunication Press, 1989.

[21] Boisot, M., "Institutionalizing the Labor Theory of Value: Some Obstacles to the Reform of State-Owned Enterprises in China and Vietnam," *Organization Studies*, Vol. 17, 1996, pp. 909–928.

[22] Xu, Y., D. C. Pitt, and N. Levine, "Competition Without Privatization: The Chinese Path," in *Telecommunication and Socio-Economic Development*, S. Macdonald and G. Madden (eds.), London, Engand: Elsevier Science, 1998, pp. 375–392.

[23] Byrd, W. A., "Contractual Responsibility Systems in Chinese State-Owned Industry: A Preliminary Assessment," in *The Changing Nature of Management in China*, N. Campbell, S.R.F. Plasschaert, and D. H. Brown (eds.), London, Engand: Frances Pinter, 1991, pp. 7–35.

[24] People's Posts and Telecommunications Newsletter, January 14, 1993.

[25] Ministry of the Posts and Telecommunications, *Brochure on China Posts and Telecommunications*, 1994.

[26] Guo, R. C., and Y. Xu, *Business Accounting Systems for Posts and Telecommunications Enterprise*, Beijing, China: Beijing University of Posts and Telecommunications Press, 1992.

[27] Xu, Y., and R. C. Guo, "Economic Reckoning Within Telecommunications Networks," *Proc. ISTN-92 International Seminar on Teletraffic and Networks*, Bejing, China, September 9–12, 1992, pp. 105–107.

[28] Blackman, C. R., "World Privatization of Telecommunications," *Telecommunications Policy*, Vol. 16, No. 9, 1992, p. 698.

[29] Wu, C. G., and X. Zhang, "An Analysis of the Seemingly High Profit in the Industry," *Posts and Telecommunications Economy*, Vol. 18, No. 1, 1992, pp. 6–9.

[30] Ding, L., "The Management of China's Telecommunications Policy," *Telecommunications Policy*, Vol. 18, No. 3, 1994, pp. 195–205.

[31] Zhao, L. X., and B. Zhu, "The State Council Affirmed the Achievement in Posts and Telecommunications, Requested for Further Rapid Development," *People's Posts and Telecommunications Newsletter*, February 16, 1993.

[32] Chen, Y. Q., "Driving Forces Behind China's Explosive Telecommunications Growth—Change in Policies Fuels Growth," *IEEE Communications Magazine*, Vol. 31, No. 7, 1993, pp. 20–22.

[33] Barney, J., "Firm Resources and Sustained Competitive Advantage," *J. Management*, Vol. 17, 1991, pp. 99–120.

[34] Fama, E. F., "Agency Problems and the Theory of the Firm," *J. Political Economy*, Vol. 88, 1980, pp. 288–307.

[35] Hoskisson, R. E., C. W. Hill, and H. Kim, "The Multidivisional Structure: Organizational Fossil or Source of Value?" *J. Management*, Vol. 19, 1993, pp. 269–298.

[36] Granick, D., *Chinese State Enterprises*, Chicago, IL: University of Chicago Press, 1990.

[37] Kornai, J., *The Socialist System: The Political Economy of Communism*, Princeton, NJ: Princeton University Press, 1992.

[38] Niskanen, W. A., *Bureaucracy and Public Economics*, London, Engand: Edward Elgar, 1994.

[39] Lindblom, C. E., "The Science of Muddling Through," *Public Administration Review*, Vol. 19, 1959, pp. 78–88.

3

Telecommunications Liberalization in 1994: Was the Time Ripe?

Compared to reforms in network development and enterprise management, reforms of institutional structure and the regulatory system had been very slow before the early 1990s. Rising pressures for liberalization, however, began to increase in intensity, mainly from private network operators and their sponsoring governmental ministries. The mid 1990s were destined to give rise to an era of managed competition.

3.1 Private Networks and Their Effects on Liberalization

The PLA and the Ministry of Railways (MOR) have traditionally constructed their own private networks for the conduct of their internal communications business. Since the late 1970s, the contrast between the rapidly growing demand for telecommunications from many ministries and the relatively insufficient capacity and poor facilities owned by the MPT became increasingly evident. As a result, many ministries followed the example of the military and railways and constructed private networks for their own internal use. The Ministries of Electrical Power, Petroleum, and Coal, as well as banks and airlines, were key examples of this phenomenon. Ironically, the centralist character of the Chinese state encouraged a system of pluralistic agency sponsorship under which individual ministries solicited sponsorship from key influentials on the State Council. Using positional power (they

were key focal ministries at the commanding heights of the Chinese economy) they habitually obtained permission to establish such private networks
in a relatively untroubled manner.

The MPT, fearful of the disaggregative effects of such private networks,
represented their inception as arising directly from the capital famine to
which it had been subjected by the government. The MPT aggressively
insisted that it could provide for the needs of such large users directly. Evidence of the salience of private networks for the MPT was recognized in
1988 with the establishment of a Communications Department. The function of this department was to manage the approval of private networks, their
coordination with the public network, and the issue of licenses for approval
of equipment connection to the public network. As Tan [1] has indicated,
the policy adopted by this department (reflective of the general policy of the
MPT) can be summed up as follows:

> [N]ew private networks will be approved only when the MPT is con-
> vinced that the MPT's public networks are not able to meet the require-
> ments of the users. Private users should first approach the MPT's public
> network for its services instead of constructing their own private
> networks.

Furthermore, it was also clearly stated by the MPT that private networks were allowed to serve other users only on condition that the MPT's
public network was not available and the deal was approved by the MPT
itself. Such muscle flexing on the part of the MPT coupled with its monopoly position may have given encouragement to the view that, in public policy
terms, the Chinese system was characterized by immobilism. The apparent
strength of the MPT position may have supported the view that in terms of
telecommunications liberalization China was a "blocked system" [2] with its
political and administrative networks effectively incapable of adapting to the
pressures of change.

Superficially, a telecommunications policy with Chinese characteristics, based upon a system of bureaucratic governance, underpinned by the
State Council and supported by the MPT, appeared to be unassailable. Yet,
there were unmistakable signs of subterranean public policy changes already
well underway. First, the power position of MPT was by no means as secure
as was sometimes maintained. There was clear indication that the operators
of private networks may have more powerful sponsors on the State Council
than the MPT itself. Thus, China's domestic airline, CAAC, was able to sign
a contract with AT&T in early 1993 to set up its own VSAT-based national

private network. In spite of the fact that strong arguments could be put forward that this network might be unnecessary (simply duplicating facilities already available through the MPT), the Communications Department was entirely unable to prevent CAAC from obtaining approval for its private network from the State Council [1].

Second, ministries owning private networks were typically cash-rich in comparison with the MPT because they possessed monopolized, well-developed infrastructure services like electricity and railways. Not only did they possess private networks with the potential for nationwide coverage, but in terms of technology such networks were feature-rich. Such private networks were—and remain—heavily digitized and based on optical fiber and satellite transmission systems. In 1993, among these private networks there were roughly five million exchange lines serving users in the railway, electricity, highway, airline, coal, petroleum, and other sectors. With these advantages, these ministries clearly found it difficult to resist the lure of the most profitable and undeveloped market segments. Accordingly, they appealed for the liberalization of the telecommunications market. In 1992, the Ministry of Electronic Industry (MEI), in conjunction with the Ministries of Railways, Electrical Power, and the PLA, put forward a proposal to form a United Telecommunications Corporation, which could compete head-to-head with the MPT. The strategy chosen was aimed at utilizing the existing private networks of these latter two ministries and the products and technology of the MEI—the major manufacturer of electronic equipment in many fields including telecommunications—to provide telecommunications services for the public.

As expected, the MPT tried by a variety of means to defend its monopoly status. Its defensive arguments, as described by He [3], can be summarized as follows:

1. Telecommunications exhibits obvious economies of scale. National resources cannot be fully exploited unless on condition that the planning and construction of the public network are unified, otherwise there would be unnecessary duplication of facilities. As a country with a telephone penetration rate of only 1% to 2% and poor long-distance facilities, China should concentrate its limited technical capacity and investment on the improvement of telecommunications under unified administrative control.

2. An effective telecommunications system could only be achieved by an integrated local and long-distance network with agreed-upon technological and service standards. Competition from private

networks will make an integrated, efficient, and reliable national public network even more difficult to achieve from the point of view of technical standards and network planning.

3. Communications should be a universal service. The ministry responsible for providing communication services has an obligation to provide services to the society as a whole, in order to promote the development of the national economy and to coordinate societal growth. Accounting policies that make cross-subsidies possible can ensure that services are extended to the remote border provinces and to the majority of the rural areas. Allowing private network operators to enter the market freely would lead to competition only in the most profitable areas or routes and in some value added services.

4. The telecommunications infrastructure constitutes the country's central nervous system and thus has implications for national sovereignty and security. A single operator makes it easier to guarantee sovereignty and security.

5. Unified planning and management of the basic network will be extremely beneficial for promoting rapid growth during the early development stage of telecommunications in China. This is why 148 of the world's nations continued to rely on a monopoly for provision of basic telephone service.

Despite the strong arguments of the MPT in support of its monopoly status, other ministries had struggled endlessly to be allowed to enter the telecommunications market. One such, the MEI, has long been a leading proponent of regulatory change. The past decade witnessed a reversal in the fortunes of this prominent telecommunications policy player.

The MEI operates 106 factories and has traditionally been the most significant manufacturer of electronic components, computers, and telecommunications equipments in China. Operators of private networks were mainly its traditional customers. With large numbers of research grants from the government and huge orders from the traditional private networks, the MEI was able to invest in more advanced R&D and technically efficient production facilities than those possessed by the MPT. For example, it was the MEI, not the MPT, that constructed China's first satellite communication Earth station in 1972.

The MEI, however, has progressively been excluded from the increasingly expanding public telecommunications market due to the MPT's closer organizational partnership with its own 29 subsidiary factories. The MPT preferred, or was obliged, to purchase equipment from its affiliated equipment manufacturing branches instead of those under the MEI. Simultaneously, a steady decline in military budgets for telecommunications has adversely affected MEI revenues since 1978. Finally, almost all private network operators, seeking to emulate the technical progress made by overseas competitors, switched procurement from their traditional supplier, the MEI, in favor of technically superior, but imported, equipment. As one commentator has stressed "the MEI's role in telecommunications has been constantly shrinking while other ministries or industries are expanding quickly in the wake of economic reform and development" [1]. Experiencing this shrinkage in its traditional markets, MEI became an ardent advocate of a more liberalized telecommunications policy.

In fact, this ministry began to adopt aggressive strategies to enter the telecommunications market as early as the beginning of the 1990s. To consummate this policy, it proposed a set of very ambitious plans, called Golden Projects, in August 1993. These Golden Projects initially consisted of four individual schemes. The first was the Golden Bridge Project, which was to build a national economic information network of stored and real-time data linking state economic departments. The second was the Golden Card Project, which aimed to create a bank credit card payment and authentication network. The third was the Golden Gate Project, an electronic data interchange (EDI) network linking all customs offices. The fourth, the Golden Sea Project, was a dedicated network for the elite leadership of the Chinese Communist Party who typically live and work in Zhong Nan Hai (the name stands for Central and South Sea). Several new golden projects have been added in succeeding years. In the light of the importance of these developments for the national economy and, thus, the strong support from the State Council, the MPT had no alternative but to cooperate.

Paradoxically, another driving force for market liberalization rests with the military, cast in the role of defender of the state's integrity. The PLA is a powerful private network operator. For reasons of state security, the army cannot open its networks for public use. According to the joint statement (order 128) issued by the State Council and the Central Military Commission in September 1993, however, responsibilities for nonmilitary and military allocations of radio spectrum were to be shared between the State Radio Regulation Commission and the PLA. It affirmed PLA authority over much of the 800-MHz frequency range. Significantly, the frequency spectrum for cellular mobile telephony partly resides within this range in accordance with international standards. So, once the market is open, the PLA could use its privilege of frequency allocation to enter the mobile communications market. The fact that certain frequency spectra in this range are freed with the help of the technique of spectrum splitting and signal compression gives the PLA more frequency resources to operate its own wireless network and to provide services to the public. Another factor that drives the military to challenge the MPT is the financial need to raise funds to cover overheads, such as military pensions, payment, uniforms, and food. In the mid 1990s, two PLA corporations made their initial appearance in the telecommunications sector. One was the Chinese Electronic Systems Engineering Corporation, which was a system integrator of telecommunications equipment. Another was the Great Wall Corporation through which most of the commercial mobile radio businesses using the CDMA network that was owned by the military were organized [4].

With the existence of these private network operators and telecommunications equipment manufacturers who can benefit from telecommunications liberalization, the effort to open the telecommunications market for competition was invigorated. The main points proselytized by the MEI and other ministries, as summarized by Fei-chang He, director of the telecommunications division of the MEI, focused on the following questions [3]:

> Which encourages more rapid development—following the beaten track or generating innovation? Which makes the fuller use of national resources—a single operator or an open system that brings every possible player into the market, especially the private networks? Which alternative better safeguards the users' interests—an exclusive monopoly or multiple carriers offering them a choice?

Corresponding to the points raised by the MPT (referred to previously), these ministries put forward the following views [3]:

1. Economies of scale are not something abstract, but must be evaluated in terms of relative labor productivity and service quality. Benchmarking with foreign countries reveals that the efficiency of telecommunication supply in China is among the lowest in the world and service quality is not up to international standards. Why are the benefits of economies of scale seemingly so elusive in the Chinese context?

2. With regard to the reasonable use of national resources, the considerable potential of private networks (especially long-distance communication capacity) remains underutilized. Thus, the existing

state-dominated system of provision was productive of negative economic consequences.

3. There can be little dispute about the value of an integrated network, but this issue should not be confused with the issue of monopoly versus competition. International telecommunications is progressing towards integration, towards "one globe, one network," yet each country has different operators. They are interconnected and made compatible according to standards and agreements enunciated by the International Telecommunication Union and other international organizations. Multiple service providers can embody all the features of an integrated network without detracting from interoperability.

4. With regard to universal service, China's private networks possess unique capabilities. The regions not covered by the public network are often typically the regions where private networks are flourishing. For example, the coverage of the railway's communication network complements that of the MPT public network; hence, utilization of private networks would be beneficial for the development of remote border provinces and rural communication.

5. The real meaning of national communications security and sovereignty is to prevent the betrayal of national secrets and to ensure that control of China's communication markets resides with its domestic carriers. The first of these goals can be achieved by strict regulation, such as censorship, while the second can be achieved by restricting the management of telecommunication services to Chinese nationals. Neither goal requires a monopoly as such. Many other countries have introduced multiple carriers without any loss of vital or sensitive information.

6. The clear international trend is away from monopoly. Since 1984, the year of the AT&T divestiture in the United States, a growing number of countries have incorporated competitive mechanisms into their telecommunications industry, which is managed, unified, and led by government-created regulators. Such a trend would appear to be inexorable.

Using such rationales as bargaining tools, the MEI and other ministries have strenuously fought for the right to enter the telecommunications market. As a result of such efforts, the MPT was forced to partially open the market.

3.2 Telecommunications Market Liberalization: The Establishment of Ji Tong and China Unicom

Under pressure from private operators and the State Council, the MPT finally issued its "Provisions on the Administration of the Liberalized Tele-communications Services" in 1993, which was validated in November of that year. Under this regulation, the MPT for the first time announced its intention to open up the operation of nine telecommunications services to non-MPT enterprises on Chinese territory:

- Radio paging;
- The 800-MHZ trunked telephone service;
- The 450-MHZ radio mobile communications service;
- Domestic very small aperture terminal (VSAT) service;
- Telephone information services;
- Computer information services;
- Electronic mail (e-mail);
- EDI;
- Videotex.

According to the arrangement, the first four services were to be subject to a licensing system, while a declaratory system was applicable to the remaining five services. This arrangement specified in detail the conditions and procedures for issuing licenses. Compared with the licensing system, the declaratory system was relatively simple, as no examination and assessment of the application was required. Written notification of whether the application has been approved or declined was required to advise the applicants within 30 days after receiving all the documents required for the application. The arrangement also stipulated that all MPT operating enterprises should provide the authorized new entrants with such relay equipment and interconnections as are essential for the operation of the services concerned in accordance with the principle of reimbursable used and mutual benefit, and to charge for them at set prices [5].

The "Provisions on the Administration of the Liberalized Telecommunications Services" was the first regulatory document regarding telecommunications liberalization in China. Since its publication, extensive competition in liberalized services has developed rapidly, especially in radio paging, as it requires relatively low investment and market demand was huge. In Beijing,

163 new radio paging service operators had emerged by the end of November 1994, and the number of subscribers jumped from several thousand to three-quarters of a million. As a result of competition, the price of radio paging has reduced sharply and service quality has been greatly improved. Pagers including the first year service charge were advertised in high street shops for 600 yuan ($69), at the end of 1994, compared with 2,000 to 3,000 yuan ($232 to $348), before the market was liberalized in 1993 [6]. For the first time, both the new operating enterprises and the customers were able to enjoy the benefits that competition could confer.

Also in 1993, the State Council authorized the establishment of the Ji Tong Communications Corporation and the China United Telecommunications Corporation (China Unicom). In June 1993, the Ji Tong Communication Corporation formally announced its registration as a company. It was organized under the MEI with stakeholders from 30 state-owned enterprises and research institutes in Beijing, Guangzhou, Shanghai, and Shenzhen. Another stakeholder is the China International Trust and Investment Corporation (CITIC), the investment arm of the State Council.

The aim of establishing Ji Tong included the articulation of joint ventures with overseas companies in communication technology and product development, and running public data and value enhanced network services in China. Ji Tong's most notable task was to undertake all the Golden Projects for MEI. Although these networks were planned for special business sectors, such as banking, taxation, and customs declaration, they have been one of the key levers for diverting value-enhanced-service traffic away from MPT. For example, in 1999, Ji Tong became the first operator in China to launch Internet Protocol (IP) telephony calling-card services with the help of its backbone data communication network. Additionally, Ji Tong provided the MEI with an opportunity "to link MEI's equipment manufacturing capabilities to the construction of an information superhighway in China" [7]. Obviously, the establishment of Ji Tong was a strategic victory for the MEI in penetrating the lucrative telecommunications market.

On July 19, 1994, the China United Telecommunications Corporation (China Unicom) was formally established. It was a joint venture with stakeholders from the MEI, the MOR, the Ministry of Electrical Power (MEP), and 13 other dominant state-owned corporations. Each of these three ministries invested 100 million yuan ($12.12 million), and each of the 13 corporations invested 80 million yuan ($9.7 million). The registered assets for China Unicom amounted to 1 billion yuan ($121.2 million).

China Unicom was founded in accordance with the Company Law of the People's Republic of China and is a limited liability company. It is a legal economic entity that is responsible for its own management decisions, and profits and losses. It prepares its own independent financial accounts. A board of directors constitutes the final decision-making authority, while the Chairman of the Board is the corporation's legal representative. China Unicom was affiliated with the State Economic and Trade Commission when it was established. Its corporate strategy—construction of infrastructure, technical upgrading, and introduction of foreign capital—was incorporated in line with plans drawn up, respectively, by the State Planning Commission, the State Economic and Trade Commission, and the Bank of China. The Administrative Office of the State Council and the Ministry of Labor directly controlled finance, personnel, foreign affairs, and salaries.

In the directive issued by the State Council, the main approved business activities of China Unicom are as follows:

- To upgrade and renovate the existing private telecommunications networks of the MOR and the MEP in order to provide long-distance telephone services to the public in addition to meeting these ministries' internal needs; and to provide local telephone services to the public to those places where the public telephone networks cannot reach or there exists a severe shortage of telephone capacity;

- To operate radio telecommunications services (including mobile phone service);

- To operate value-enhanced services;

- To undertake engineering projects for all kinds of telecommunications systems;

- To conduct other services related to its main business activities.

The telecommunications operation of China Unicom was, however, subject to the MPT's regulation. According to the State Council, China Unicom's telecommunications networks, either upgraded or newly constructed, were eligible to be connected to the public telecommunications network through automatic interfaces. Fair settlement should be arranged to achieve an agreed common share of telecommunications resources, such as fair payment for network interconnection.

With the Chinese government attaching increasing importance to the development of telecommunications (as suggested in Chapter 2), it conferred favorable operating freedoms on the MPT. These included low tax liability and high depreciation rates. The State Council was to bestow similar conditions on China Unicom, thus ensuring greater operating equity and the construction of a more level competitive playing field.

In his speech at the opening ceremony of the establishment of China Unicom, Zou Jia-hua, the vice premier, pointed out strongly the importance of establishing China Unicom as follows [8]:

> The development of telecommunications depends upon huge amounts of investment. The establishment of China Unicom will encourage other industrial sectors to invest in telecommunications, so the overall financial resources are expanded. Another purpose enshrined in the establishment of China Unicom is the creation of a mechanism for co-operation and competition in the telecommunications sector and the breaking-up of the monopoly, in the interests of positively improving service quality, enterprise management and technological innovation.

He also urged that the management of China Unicom should be conducted along the lines of a modern enterprise system, with clearly defined asset ownership, separated governmental administration, and business operation functions. At the same time, he declared that in order to foster success in this experiment, the provision of basic telecommunication services to the public by any other corporations without the permission of the State Council would be strictly banned [8]. In this way, China Unicom could concentrate its limited resource on competing solely with the incumbent. In the meantime, the declaration also indicates that the government has taken a cautious, step-by-step approach in liberalizing the telecommunications market.

The establishment of China Unicom represented a milestone in the development of telecommunications in China and marked the end of the MPT's monopoly in telecommunications operations and the beginning of open competition in telecommunications services, including basic telephony. At the same time, it raised expectations among foreign investors of greater opportunities for participation in the Chinese telecommunications market. Such opportunities had been effectively stymied in the past.

3.3 The Market for Foreign Investors

The primary form of foreign involvement has been in the telecommunications equipment field. Since 1978, the demand for telecommunications increased dramatically while the poorly equipped telecommunications system in China could not effectively respond. The government recognized the increasing technological gap between the Chinese telecommunications system and that of the developed countries. It became increasingly clear that to wait for domestic R&D to make breakthroughs and develop to the same level as more advanced countries was a forlorn cause. In such circumstances, the door for the import of foreign telecommunications equipment was prized open.

The telephone switching system of the Chinese public telephone network was traditionally based on step-by-step and cross-bar technology. In 1982, the Fuzhou telecommunications office installed the first computer-programmed telephone switching system [Stored Program Control (SPC)] in China, which was imported from the Fujitsu Corporation of Japan. The huge capacity, high connection speed, and excellent quality of this system were very impressive and attractive to operating enterprises in other parts of China. Consequently, this was quickly followed by a wave of other imported Stored-Program-Control, or SPC, telephone switching systems throughout the country. Other areas of telecommunications import penetration included digital microwave transmission systems, satellite Earth station systems, and cellular mobile systems.

As for customer premises equipment (CPE), the market has been fully opened provided that products meet international technical standards and the particular requirements of the Chinese network. Manufacturers, however, have to apply for a license from the regulator. As CPE is not too technologically complicated, this market has been extremely competitive.

Local governments sponsored most of the foreign currencies used by local telecommunications offices for importing foreign equipment, as they have realized how important an advanced telecommunications system is to the development of the local economy. At the same time, soft loans from foreign governments have also been important sources of foreign currency. Table 3.1 shows soft loans to China in the telecommunications industry between 1982 and 1989. Normally, these soft loans were designated by foreign governments for the purchase of telecommunications equipment from their domestic vendors. In consequence, "the companies who have claimed important stakes in the Chinese telecommunications market are coincidentally those whose governments are leaders in providing concessional loans to China" [9]. For example, during the late 1980s, Alcatel of France, NEC of Japan, and Siemens of Germany were three leading suppliers in the field of central office switching [10].

Almost all the major international equipment suppliers have representative offices in China, including Alcatel, AT&T, Ericsson, Northern Telecom, Siemens, Motorola, and Nokia. Some of them have established very close cooperative relationships with the Chinese telecommunications sector

Table 3.1
Soft Loans to China in the Telecommunications Sector (1982–1989)

Country	Loans ($ million)
France	169.0
Japan	135.0
Canada	116.0
Sweden	85.7
Spain	35.0
Germany	30.0
Netherlands	29.7
Belgium	21.0
Italy	15.0
Norway	14.8
Total	645.5

Source: [9].

through the establishment of joint ventures in equipment manufacturing. One of the most significant examples is the agreement between the China Posts and Telecommunications Industry Corporation (PTIC) and Alcatel's branch in Belgium in 1983. According to this agreement, a two-stage technology transfer process was initiated. During the first stage commencing in 1984, a joint venture, under the name of the Shanghai Bell Telephone Equipment Manufacturing Company, would install more than 100,000 lines of Alcatel's S1240 switching systems in Shanghai, Beijing, and Tianjin. During the closing period of stage one in 1986, the venture would build a factory where the switching system could be assembled with parts imported from Belgium. In the second stage, full technological and manufacturing capability would be transferred to the venture. This would enable the factory to produce the S1240 system independently from parts supplied by Alcatel. More importantly, it also included the technology and manufacturing capacity for wafer fabrication as well as testing and packaging of the custom microprocessor at the switch's heart. Experience has shown that this joint venture has been very successful in China. It has continuously ranked among the top 10 most profitable joint ventures in China for many years [11].

Another example of a successful international player in the Chinese market is the equipment supply branch of the former AT&T (currently

Lucent Technologies). Regretting that it was a latecomer in the Chinese tele-communications market, AT&T subsequently intensified its efforts to compete with other suppliers from October 1985 when it opened its first corporate representative office in Beijing. It was the first foreign corporation to sign a comprehensive Memorandum of Understanding with the State Planning Commission of China in February 1993. Although this was not a sale contract, it guaranteed long-term cooperation between the former AT&T and the Chinese telecommunications industry. It defined 10 areas of further cooperation, including 5ESS(R) switch manufacturing, VLSI manufacturing, network management expertise, R&D with Bell Labs, optical transmission manufacturing, wireless manufacturing, customer premises equipment manufacturing, training, system integration, and network service offerings [11]. In February 1995, AT&T established a completely self-owned subsidiary, AT&T China Company, which has acted as an administrative center for overseeing the entire portfolio of AT&T activities in China [12].

There are clear foreign ambitions to enter not only the equipment market but also the public telecommunications network and services market segment, as this sector is of even greater potential. An early trial involving foreign telecommunications firms involved the offering of call-back service in China. To use this service, a resident of China would register his or her phone number with an overseas company. Then, when the user wished to place an international call, he or she would call the company, but only allow the phone to ring a few seconds. Then the company would call the user back according to the caller's identity code and the user inform the company of the number with which he wished to be connected. Since the call originated from foreign liberalized markets, the tariff rate is relatively lower and the customer can save money. In this way, operators can make profits by exploiting the rate difference between MPT and key foreign firms, thereby competing with MPT through by-passing MPT's international network [13]. Perhaps unsurprisingly, this service met with a vigorous response from the MPT. In May 1995, the Department of Telecommunications Administration of the MPT issued a tersely worded announcement banning the resale business including call-back services. The MPT announcement read as follows [14]:

> According to the policy of the Chinese government, international tele-communication service (in China) can only be provided by the MPT. No individual or organization, including the reseller, can participate in international telecommunications service provision in any form. Recently, we found that certain foreign companies are advertising in China to provide call-back service. This is a severe violation of China's regulations regarding international telecommunications services. We

solemnly announce that all resale of international service must stop. Users of call back service should stop using that service immediately. Otherwise, we will take necessary measures.

The MPT intimated that penalties for violation would comprise warnings, fines, and eventual service line cut-off [15].

The MPT has successfully prevented foreign direct ownership of the network and foreign direct involvement in its service operations. One of the rationales that was paraded to defend this policy, according to Wu Ji-chuan, Minister of the former Ministry of Posts and Telecommunications and the current MII, was the argument that the "posts and telecommunications service is closely connected with the country's political and social activities. It is, therefore, related to the sovereign right of the state, and has to be centrally controlled [by the Chinese government]" [16]. This kind of concern is understandable. As was pointed out in Chapter 2, the Chinese retain unfortunate memories of the early days when foreign operators dominated the Chinese telecommunications system. Mueller [17] has pointed out the following:

China has a centuries-long tradition of xenophobia, reinforced by a century and a half of mistreatment and subjugation by Western powers, yet its growth is driven by foreign technology and investment, and is highly dependent on exports to foreign countries. China routinely trades access to its market for foreign technology and investment while guarding its sovereignty and autonomy with a ferocious zeal. Its leadership must simultaneously lure in foreigners and hold them at bay.

A subtle reason for the MPT defending its closed-door policy was that the MPT, as the incumbent ministry, has long recognized the market potential of telecommunications against a background of double-digit economic growth. A "twin track policy of exclusivity" has led it to repel what it perceives as cream skimming on the part of foreigners, as well as domestic competitors. The earliest example in the 1980s was the formation of the Shenda Telephone Corporation, a joint venture between the Shenzhen Posts and Telecommunications Administration and the former Hong Kong Telecom, a branch of Cable & Wireless in Hong Kong. It provided a golden opportunity for penetrating the Chinese telecommunications operations market, as Shenzhen was the first special economic zone in China, which encouraged the adoption of flexible provisioning policies and experiments. The joint venture, however, came to a premature end as Cable & Wireless's stake was bought back by Guangdong Posts and Telecommunications Authority due to the unimaginably high profits this corporation was earning. This unfortunate debacle discouraged the process of liberalization in China, and this early-bird entrant has been seriously criticized by individuals defending the closed-door status of the MPT. For example, Gao [18] argued that "in the development of [telecommunications in] the Shenzhen Special Economic Zone, we have already learned a lesson paid with losses, which we should keep in mind."

Despite the MPT's strict restrictions on foreign direct investment, there still emerged some grey and flexible areas with the increasing autonomy of local PTAs. The earliest indication of the presence of such areas can be found in the MPT's response to the news reported by a Hong Kong–based newspaper that some companies in Hong Kong had signed agreements with certain mainland companies to jointly manage China's telecommunications services. An MPT spokesman reaffirmed in May 1993 that the government would continue to retain its full control over the management of China's telecommunications system. He inferred that the Chinese government would continue its decades-old policy of barring nonmainland companies and individuals from involvement in the operation of the country's telecommunication services, or from becoming shareholders in China's telecommunications system—both wired and wireless. He did not confirm whether these agreements had been signed or not, but demanded immediate stoppage of violations of government policy [19]. This speech constituted not only a declaration of the MPT in favor of domestic domination, but also a warning to local PTAs, although the spokesman did not indicate what penalties would be imposed for defiance of this decree.

One month later, the Ji Tong Communication Corporation was formally established. This created another grey area, as Ji Tong did not belong to the MPT. This enabled Ji Tong to act as a backdoor for foreign operators and a bypass agent. In September 1993, Ji Tong announced that it had set up a joint venture with Pacific Link Communications Limited of Hong Kong in the hope of developing the mobile phone market on the mainland. The 30-year-term joint venture—Beijing Ji Tong Pacific Link—would work on communication network design [20]. Although this joint venture has not directly participated in the network operation, it indicates that foreign corporations have made a further inroad into the Chinese telecommunications network and services market.

The establishment of China Unicom further widened the backdoor for foreign involvement. According to Zhao Wei-chen, the founding Director General of China Unicom, a significant proportion of China Unicom's investment was to be raised through "bold experiments" in foreign financing, although he declined to elaborate [21, 22]. Experience since China Unicom's establishment has shown that the company has played an increasingly important role in attracting foreign capital, and that the MPT has quietly changed its attitude to foreign investment. This transition will be reviewed in later chapters.

3.4   From Monopoly to Competition: Was the Time Ripe?

The global trend towards liberalization and deregulation, the existence of well-equipped private networks, ambitious domestic equipment suppliers, and aggressive foreign corporations, together with an increasingly high public demand for telecommunications services, placed the Chinese government under intense pressure to liberalize the indigenous telecommunications market, which until 1994 was completely monopolized by the MPT. With the establishment of Ji Tong and China Unicom, telecommunications liberalization was finally underway in China.

When China Unicom was established in 1994, however, telephone mainlines per 100 inhabitants were only 1.45, which was significantly lower than that of early-mover countries when their telecommunications markets were liberalized. For example, when the Bell System split up in 1984, teledensity was 46% in the United States. Consideration of the effects of the U.S. case raised serious concerns over the timing of Chinese telecommunications market liberalization, (i.e., the question arose whether termination of the monopoly operation of the Ministry of Posts and Telecommunications was, in fact, premature) [23].

Pitt et al. [24] itemized three generic factors that have played a key role in the process of telecommunications liberalization on a worldwide basis. The presence of these factors in any country arguably provides the necessary preconditions for the existence of telecommunications competition and promising opportunities for new entrants.

The first factor is firmly rooted in the area of technological change. Although telecommunications in China developed very late from the beginning of the 1980s, it benefited considerably from the convergence of the techniques of computing and telecommunications, which appeared and were applied in the late 1970s and early 1980s. With late-mover advantage, the Chinese telecommunications system has developed from a very low technical base to one of the most technically advanced telecommunications systems in the world. According to the indicators database of the International Telecommunications Union, 96.5% of the MPT's mainlines were digitized by 1994. This figure was even higher than that of the United States (71.6%)

and the United Kingdom (82.7%)—the two telecommunications liberalization leaders—in the same year.

According to Pitt et al. [24], technological modernization of a telecommunications system can strongly influence telecommunications liberalization and deregulation. On the one hand, digitalization makes it effective and economic for networks owned by different operators to be connected to each other. Therefore, there should not be any technical and economic barriers for China Unicom to connect its own network to that of China Telecom. On the other hand, modernization provides strong stimulation to demands for a full range of innovative electronic communications services including broadband multimedia and high-speed Internet, which in turn require an efficient and competitive telecommunications market. It will be very difficult for a bureaucratic monopoly to effectively manage all of these services satisfactorily, and pressures for a liberalized telecommunications market will undoubtedly make an appearance [25]. Technical changes in the Chinese telecommunications system not only demonstrate the feasibility of telecommunications competition, but also underline its necessity.

The second impeller of change is the influence of the telecommunications liberalization experience of other countries and the policy transfer of such ideas into the Chinese politicoeconomic milieu. The privatization of British Telecom and the divestiture of AT&T marked the beginning of the contemporary telecommunications liberalization process. The encouraging success of these two experiments stimulated telecommunications liberalization in other countries, especially in Europe. This process, undoubtedly closely examined by the Chinese government, had a major impact on its thinking. When the State Council received the joint proposal from the MEI, the MOR, the MEP, and the PLA for the establishment of China Unicom, it quickly formed a special group to study the feasibility of liberalizing Chinese telecommunications [unattributed interview at the Institutional Reform Research Center of the State Council, October 14, 1994]. This group conducted a series of foreign visits and produced a thorough review of the telecommunications liberalization process of these first-mover countries, especially that of the United States. Based upon its study, the State Council agreed upon a joint proposal at the end of 1993, but the PLA was still not allowed to join with this corporation for reasons of national security. The experience of other countries has rendered telecommunications liberalization and deregulation an acceptable idea in China.

The third influence is pressure from large business users. As China moves closer towards an information society, more and more business users become dependent on a seamless and sophisticated information infrastructure for the effective management of their businesses. Such large business users as banks and customs offices have demanded both effective and economic communication services, which cannot be satisfactorily provided by the MPT. The four Golden Projects mentioned above were initiatives of the MEI to respond to their demand, which obtained strong support from the central government. In this case, these large business users acted as an accelerator of telecommunications liberalization in China.

In addition to the above factors, there was also a strong economic propellant for liberalization of the Chinese telecommunications industry circa 1994. In fact, in early 1994, the Ministry of Posts and Telecommunications formally announced that the telecommunications infrastructure in China was finally able to satisfy the basic demands of the public and the economy. The waiting list for installing telephone services, which was as long as 3 years in major cities in the 1980s, no longer existed. This was a critical turning point: The Chinese telecommunications market had turned from a sellers' into a buyers' market [23]. Installing a telephone via the backdoor or via personal connections with staff of telecommunication offices, finally became history, and extensive promotion of telephone line installation was conducted by all provincial telecommunications enterprises.

The Chinese government applauded this achievement and, at the same time, realized that the time was ripe to transform the Chinese telecommunications sector into a market-oriented industry. Operational efficiency became the top priority on the government's agenda, as it clearly realized that high telecommunications growth rates in the past had mainly resulted from preferential policies and significant investment. Figure 3.1 shows total telecommunications investment as a percentage of overall GDP in selected countries. It clearly indicates that China has given increasingly higher priority to public telecommunications investment since 1980, and has reached and surpassed investment levels in other major economies.

It is true that the penetration rate in China was still much less than the penetration rate of the United States when the American telecommunications market was deregulated in 1984, but the fact was that the Chinese market had developed into a buyers' market as it did in the United States before market liberalization. To a great extent, the Chinese telecommunications sector in the early 1990s was facing the same challenges as those faced by early-mover countries in the early 1980s, namely pressure for improving efficiency. Development that was overreliant on government's preferential support and heavy investment would not be sustainable. This implied that there was an immediate and irresistible urgency to liberalize the Chinese telecommunications market within the context of demonstrably increasing demand.

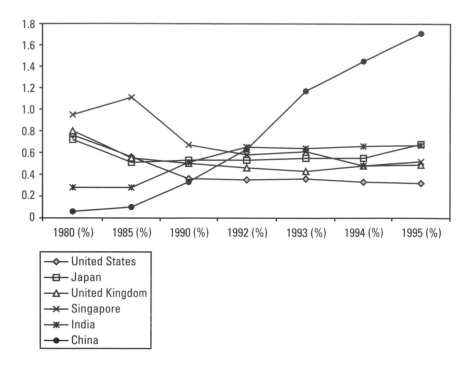

Figure 3.1 Total telecom investment as a percentage of GDP. (*Source:* ITU World Tele-communication Indicators Database.)

3.5   Conclusion

From the beginning of the 1990s, the Chinese government gradually withdrew the preferential policies it once granted to the telecommunications sector, and opted instead to liberalize the telecommunications market. The highly digitized infrastructure, the encouraging experience of telecommunications liberalization elsewhere in the world, and the pressure from large users, private network owners, and domestic equipment suppliers all contributed as propellants of this process. Importantly, the government clearly recognized that the rapid growth of telecommunications since 1978 had been mainly due to the preferential support of the government and generous investment. Such support for growth and investment-intensive development might be necessary in a situation in which expansion of telecommunications infrastructure lagged far behind the demands generated by a growing economy. But, in order to guarantee the efficient and sustainable development of the telecommunications sector, full incorporation of market mechanisms was deemed essential. Against this background, 1994 can be viewed as the *annus mirabilis* of reform. The date signals the final transition of the Chinese telecommunications sector from a sellers' to a buyers' market.

The establishment of China Unicom heralded a milestone in the Chinese government's former telecommunications policy—marking a break with the traditional centralized monopoly model [26]. The Chinese telecommunications policy path, however, has been obviously different from that of early-mover countries. In particular, a new telecommunications policy with Chinese characteristics, embracing partial or full liberalization of networks and services, may be expected to ultimately produce discrete models of liberalization and deregulation attuned to the Chinese political/economic context. Significantly, the Chinese have a particularly apposite phrase to characterize contemporary policy developments in the telecommunications area: touching stones to cross the river [24]. As in other policy sectors, Chinese government policy in telecommunications appears to resonate to the message of incrementalism, or a step-by-step approach, rather than a headlong response to the tocsin of synopsis, connoting a radical break with the past. In contrast to the recent history of Hong Kong's telecommunications sector, Chinese policy experimentation evinces gradualism and caution [27]. The incorporation into the Chinese context of avowedly Western market models derived from the tenets of economic liberalism will be suitably nuanced. The dialectic between central control and market contestability endures as a key characteristic of contemporary Chinese telecommunications policy.

Clearly, Chinese telecommunications policy mirrors developments in the wider society, and telecommunications policy outcomes reflect the huge changes emanating from the country's tumultuous twentieth-century history. It also reflects the pragmatism that has done much to temper the ideological radicalism that coursed throughout Chinese society under Mao Ze-dong. Despite huge and growing disparities of wealth and income between rich and poor and urban and rural society, China is now reaping the benefits of a move to a market form of socialism. Such may be depicted as marking a second Great Leap Forward in Chinese society at the beginning of the twenty-first century. Transformations in telecommunications provision mirror these wider changes—China's position as a second mover has, in fact, conferred upon it huge advantages and enabled it to leapfrog to a position of huge technological potential while allowing it to proceed more successfully along a transitional path than its counterparts in the former communist Eastern Europe. Chinese society, however, can be accurately characterized as a hybrid in which the tenets of state socialism, while under constant pressure from technological, social, and even political change, remain resilient, suggesting that gradualism and a managed transition will continue to hold center stage as political and administrative imperatives. The management of competing pressures of conservation and change (giving rise, in turn, to complexity and occasional dissonance in decision making) seems likely to continue as a central characteristic of public policy in the telecommunications sector in company with the rest of the Chinese political/industrial complex.

References

[1]   Tan, Z. A., "Challenges to the MPT's Monopoly," *Telecommunications Policy*, Vol. 18, No. 3, 1994, pp. 174–81.

[2]   Crozier M., *The Bureaucratic Phenomenon*, London, England: Tavistock, 1962.

[3]   He, F. C., "Lian Tong: A Quantum Leap in the Reform of China's Telecommunications," *Telecommunications Policy*, Vol. 18, No. 3, 1994, pp. 206–210.

[4]   Ure, J., "Telecommunications, with Chinese Characteristics," *Telecommunications Policy*, Vol. 18, No. 3, 1994, pp. 182–194.

[5]   Xu, S. Y., *Brochure of the Ministry of Posts and Telecommunications*, Beijing, China: Ministry of Posts and Telecommunications, 1994.

[6]   China Telecom Newsletter, December 1994.

[7]   Mueller, M., and Z. Tan, *China in the Information Age—Telecommunications and Dilemmas of Reform*, Westport, CT: Praeger, 1997.

[8]   Zou, J. H., "Speech on the Opening Ceremony of China Unicom," *People's Posts and Telecommunications Newsletter*, July 21, 1994.

[9]   Yan, M. Z., "The China-Hong Kong Relationship in Telecommunications," in *Telecommunications and Development in China*, P. Lee (ed.), Cresskill, NJ: Hampton Press, Inc., 1997 pp. 201–223.

[10] Wang, W. P., "China: Adapting to New Needs," *1992 Single Market Communications Review*, Vol. 3, No. 2, 1991, p. 68.

[11] Warwick, W., "A Review of AT&T's Business History in China: The Memorandum of Understanding in Context," *Telecommunications Policy*, Vol. 18, No. 3, 1994, pp. 265–274.

[12] Chismar, W. G., M. Jussawalia, and M. S. Snow, "U.S. Provision of Telecommunications Goods and Services in the PRC: Chinese Policies and American Strategies," *Telecommunications Policy*, Vol. 20, No. 6, 1996, pp. 455–464.

[13] China Telecom Newsletter, June 1995.

[14] *People's Daily*, May 22, 1995.

[15] Shan, J., and J. Liu, *China News Digest*, June 11, 1995.

[16] *China Daily*, August 26, 1997.

[17] Muller, M., "China: Still the Enigmatic Giant," *Telecommunications Policy*, Vol. 18, No. 3, 1994, pp. 243–253.

[18] Gao, Y. Z., "Shanghai's Telephone Network: Status Quo and Development Strategy," *Posts and Telecommunications Economy*, Vol. 16, No. 3, 1991, p. 24.

[19] *China Daily*, May 11, 1993.

[20] *China Daily*, September 22, 1993.

[21] Ingelbrecht, N., "All Aboard the China Express," *International Communications Week*, Issue 133, 1994.

[22] Ingelbrecht, N., "China to Break Open Monopoly," *International Communications Week*, Issue 109, 1994.

[23] Kan, K. L., "Where to Go for the Chinese Telecommunications Industry?" *Posts and Telecommunications Economy*, Vol. 24, No. 10, 1999, pp. 4–6

[24] Pitt, D. C., N. Levine, and Y. Xu, "Touching Stones to Cross the River: Evolving Tele-communications Policy Priorities in Contemporary China," *J. Contemporary China*, Vol. 5, No. 13, 1996, pp. 347–365.

[25] Mueller, M., *Telecom Policy and Digital Convergence*, Hong Kong, China: City University of Hong Kong Press, 1997.

[26] Xu, Y., "The Impact of the Regulatory Framework on Fixed-Mobile Interconnection Settlements: The Case of China and Hong Kong," *Telecommunications Policy*, Vol. 25, No. 7, 2001, pp. 515–553.

[27] Xu, Y., and D. C. Pitt, "One Country, Two Systems: Contrasting Approaches to Tele-communications Deregulation in Hong Kong and China," *Telecommunications Policy*, Vol. 25, Nos. 3, 4, pp. 245–260.

4

China Unicom: Victim of Reversed Regulatory Asymmetry?

The establishment of China Unicom in 1994 marked a major milestone in the development of telecommunications in China. It clearly indicated a fundamental paradigm shift in the evolution of the Chinese telecommunications system (i.e., the termination of the historical monopoly operation of the public telecommunications network and the commencement of telecommunications liberalization in China). Significantly, its business operations were largely separated from direct government intervention. This move signaled a clear commitment on the part of the country's policy makers to break with traditional bureaucratic patterns of administration and service delivery and engender economic liberalization and pluralistic provision, presaging further attempts at structural reform while simultaneously acting as the harbinger of change elsewhere in the Chinese economy.

Emergence from the womb of a bureaucratically structured provisioning system is normally an uneven process, especially for infrastructural industries, such as telecommunications, which long enjoyed monopoly status. It involves weaning an organization from safety-first habits typical of a bureaucratic decision-making organizational apparatus and encouraging the adoption of a risk-taking, proactive, and competitive strategy more typical of a business organization. The experience of China Unicom during the period from 1994 to 1998 indicates that such a journey may be hard and troublesome for a new entrant, especially in the absence of a well-developed regulatory framework.

4.1   The Telecommunications Regulatory Framework: 1994 to 1998

The MPT had been a monopoly operator of the public telecommunications network in China long before the establishment of China Unicom. It comprised several departments. The DGT was responsible for coordinating the operations and maintenance of the telecommunications network and services, while the DGP was similarly responsible for coordinating the operation of the postal service. The Department of Finance centrally managed financial affairs, while the Department of Construction managed construction and investment projects. The Department of Planning was engaged in network planning and project budgeting. The Department of Communications was responsible for coordination of the public and private networks. Other key departments included the Departments of Personnel, Policy and Law, Science and Technology, and Foreign Affairs. The overall organizational structure adhered to the bureaucratic model of the former Soviet Union in the 1960s. Coordination among different departments proved very difficult and was typically ineffective. The central government realized these problems and required the MPT to propose a plan for organizational restructuring at the beginning of 1992, immediately after Deng Xiao Ping's visit to Shenzhen where he took the opportunity to urge rapid reform in the Chinese economy. In early 1994, the State Council agreed upon the organizational restructuring plan proposed by the MPT. The reshaped organizational structure of the MPT is shown in Figure 4.1 [1].

According to this plan, the function of the MPT metamorphosed from the purely routine operation of the postal and telecommunication systems into active regulation of the postal and telecommunication markets. The functions of governmental administration and business operation have, to some extent, been separated. As shown in Figure 4.1, all branches named *department* would henceforth undertake the role of governmental administration, while the remainder would focus on business operations and business support [2].

The Department of Communications was renamed the Department of Telecommunications Administration (DTA). It was designated a significant role as the regulatory arm of the MPT for the national telecommunications sector. According to the restructuring plan, the DTA was committed to making telecommunications policy, formulating and implementing

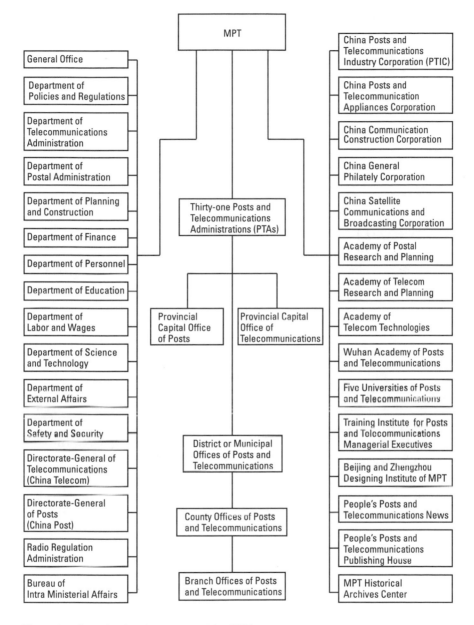

Figure 4.1 Organizational structure of the MPT.

telecommunications regulations, setting quality standards for telecommunication services, and issuing licenses.

The DGT became a commercial branch of the MPT with separate accounts and was henceforth responsible for the operation of the national public telecommunications network in addition to the provision of universal service for the public. The financial and planning responsibilities of the DGT, however, were still managed centrally by the Department of Finance as well as the Department of Planning and Construction. As a step towards separation from the MPT, the DGT officially registered with the State Administration for Industry and Commerce (SAIC) as a company called China P&T Directorate General of Telecommunications, which in short form became China Telecom in 1995 [3]. According to the license issued by the SAIC, China Telecom inherited all telecommunication services that were originally offered by the MPT. Soon after this restructuring, China Telecom physically moved out from the MPT headquarters.

At face value, this appeared to be a very impressive organizational restructuring. The DGT, or China Telecom, however, essentially remained as an administrative agency that did not actually provide any of the services itself. The nationwide telecommunications network, including the three international gateways, was actually operated by provincial PTAs under the brand name of China Telecom. The PTAs reported directly to the MPT and communicated with individual departments of the MPT, the DGT being one of the many. These PTAs enjoyed parallel administrative status with the DGT rather than remaining as its subordinate operators. In fact, some of them attempted to distance themselves from China Telecom as much as possible for the sake of autonomy. For example, the PTA in Beijing was intended for renaming as the China Telecom Beijing Branch Company, but instead, it styled itself as Beijing Telecom with a distinct corporate logo. The DGT did not own any telecommunication assets except the brand name: China Telecom. Each year the MPT, the official regulator, reportedly allocated 0.5% of its annual revenue to the DGT towards its expenditure, a figure of around 400 million yuan ($48 million) [3]. In this sense, the DGT was in fact an isolated titular headquarters of China Telecom while, in effect, enjoying very limited self-autonomy. Main operational control remained in the hands and under the jurisdiction of individual MPT departments.

According to interviews conducted with officials of the MPT, at least two reasons existed for the MPT's slow and incomplete organizational restructuring [unattributed interview, October 12, 1996]. The first related to organizational politics. The MPT was extremely reluctant to concede its controlling power over the whole network. Second, it clearly preferred a smooth transition rather than revolutionary change in the process of structural reform. This was presumably because telecommunications in China was then at its zenith—the demand for telecommunications service was historically high as a result of China's rapidly growing economy. In the light of maintaining momentum in the sector, the MPT was fearful of missing opportunities afforded by concentration on a focused market approach. Dubious schemes of internal reorganization were critically viewed within the ministry as diverting attention from this primary objective.

From 1994 to early 1998, the organizational structure of the MPT retained its vertically integrated character, while the DGT (China Telecom) remained as one of the subordinate departments of the MPT. China Unicom was supposed to compete with China Telecom but, in reality, was forced to compete instead directly with the MPT. Unsurprisingly, as noted by Gao and Lyytinen [4], the MPT adopted heavily "twisted" or reversed asymmetric forms of regulation aimed at supporting the incumbent while deterring new entrants. This strategy contrasted vividly with dominant regulatory practice in other countries where the adoption of liberalization schemes was typically accompanied by the crafting of asymmetric approaches deliberately designed to favor new entrants rather than the incumbent telecommunications provider.

Such an outcome in China suggests that the shape and pace of the telecommunications liberalization process, and the philosophy of market contestability at its heart, are contingently related to a number of key institutional characteristics (including political/ideological and cultural values), which will vary between countries. Mesher and Zajac [5], for example, have recognized that the reason for the rate of telecommunications liberalization varying widely from country to country is crucially related to patterns of rivalry between powerful interest groups involved in the telecommunications policy process. Such groups—of roughly equal political power—will typically battle to establish influence, especially in situations of fluidity and rapid change. The form and outcome of such struggles will depend on each nation's history, demographics, property rights, and legal structure. For example, as was revealed by a comparative study of telecommunications policy in Japan and the United States, differences in political structures in these two countries have led to different consequences for telecommunications users and equipment manufacturers—despite the two governments professing to have the same telecommunications policy objectives [6]. Such studies indicate that the institutional structures and values within which the telecommunications policy process is enmeshed will produce widely varying outcomes as a result of "path dependency" [6].

The above "institutionalist" argument implies that the composition and interactions of key interest groups will play a critical role in setting the direction of telecommunications liberalization and deregulation. This raises, in addition, the issue of the key factors that exert influence on the composition of individual interest groups and their patterns of interaction. In addition to Mesher and Zajac's study, Xu and Pitt [7] and Xu [8] have revealed that the shape of the regulatory framework is itself one such factor in the formulation and activities of such groups and may, in consequence, force regulators to adopt nuanced regulatory stances when dealing with issues, such as network interconnection. As a result, different forms of competition may characterize the policy paths taken by liberalization and privatization in different political, economic, and cultural settings.

In China, the close affiliation of China Telecom with the MPT has led to a situation in which, to a considerable extent, the MPT has had a direct interest in the performance of China Telecom. The interests of the incumbent operator and the regulator have been closely aligned. As a result, they constituted a unified interest group until 1998. In the absence of any legal framework or clear policy guidelines, such a situation meant that it was inevitable that new entrants would be treated unfavorably. The problematic settlement on network interconnection between China Unicom and China Telecom, which will be highlighted in the next section, indicates that the MPT had an interest in creating formidable barriers to China Unicom in its role as a competitive rival.

4.2   Interconnection Settlements Under Reversed Regulatory Asymmetry

Network interconnection has been the source of key disputes between China Unicom and the MPT as well as its operating arm (China Telecom). In an interview with China Unicom officials conducted in October 1994, China Unicom complained that the MPT and China Telecom appeared to have no intention of cooperating over interconnection. Whenever China Unicom officials approached the MPT and China Telecom, they were always treated perfunctorily and negotiations on interconnection were characteristically protracted.

Ironically, during an interview conducted at the DTA of the MPT later in the same month, a DTA official complained that visitors coming from China Unicom were usually senior officials, not engineering experts, who insisted on only meeting MPT officials at the same official level rather than staff deemed to be working in areas of low esteem with a practical as opposed to intellectual focus. The inevitable result was that the content of discussions was high level and very general. Such discussions typically failed to focus on salient issues.

The DTA suggested that China Unicom propose a technical scheme for interconnection and negotiate details directly with China Telecom. In the event of disputes arising, the DTA, as the regulatory arm of the MPT, would intervene to achieve rapprochement between the two parties. Due to the absence of published guidelines in the early stages of discussion, however, negotiation between the two operators proved to be a lengthy and drawn-out process. Frequently, the parties would end up approaching the DTA to provide a solution. The key problem emerging from this process was that it proved very difficult for China Unicom to obtain even-handed, much less enthusiastic, support from the DTA due to the close organizational relationship between the DTA and China Telecom (both under the same MPT umbrella). A DTA official informed one of the authors that China Telecom was (and remains) a "national flag carrier" in the Chinese telecommunications industry and therefore should certainly be given "most favored organization" treatment.

To obviate this kind of bias and discrimination, China Unicom had to fully utilize its relative advantage—notably the political clout of its stakeholders. Its sponsoring ministries—key players in the Chinese system of bureaucratic politics—were able to exercise a predominant influence in the State Council due to their pivotal positions as the sponsors (and effective guardians) of important industrial sectors in China. The settlement of the interconnection charge for China Unicom's mobile phone service provides an appropriate example.

Mobile services, the first type of service provided by China Unicom to the public, commenced in late 1995. Negotiations on interconnecting China Unicom's mobile networks with China Telecom's fixed and mobile networks began almost immediately after China Unicom's establishment in July 1994. In addition to a problematic technical specification on network interconnection that was issued by the MPT in June 1995, core disagreements arose on appropriate interconnection charges. Since this dispute was stalled and neither side made any concessions, China Unicom was forced to utilize its political strength and report the matter to the State Council. Tellingly, on this occasion, the State Council authorized the more neutral State Planning Commission (SPC) to deal with this issue, instead of the regulator (MPT).

With the intervention of the State Planning Commission, both China Unicom and China Telecom finally reached a compromise on interconnection. In March 1996, the State Planning Commission issued a regulatory document on the financial settlement for network interconnection [9]. The main elements of this settlement were as follows:

- China Unicom should pay China Telecom 0.08 yuan ($0.0096) for every 3-minute call originating from its mobile subscribers to China Telecom's local fixed telephone customers.

- China Telecom should pay China Unicom 0.01 yuan ($0.0012) for every 3-minute call originating from its local fixed telephone customers to China Unicom's mobile subscribers.

- No transfer payment would be made between China Telecom and China Unicom for communications between the mobile subscribers of each company (sender keeps all).

- For domestic long-distance calls, 92% of the charges should be transferred to the party that undertakes the long-distance transmission element of the call.

- All China Unicom's income for international calls should be transferred to China Telecom.

There have been subsequent arguments over the settlement of interconnection charges, as these rates were mainly based on the retail tariffs of China Telecom, which have remained unchanged for years due to the previous monopoly operation of the MPT. Costs, and much less incremental costs, were not considered as a factor when setting the interconnection charges. Thus, China Unicom argued that it was being overcharged (interview with officials of China Unicom's Department of Interconnection in June 1996). Upon closer examination, however, it seems obvious that the real disadvantages accruing to China Unicom have stemmed more from the MPT's technical specifications, rather than interconnection charges.

The MPT issued its dictum, "Technical Specifications of the Relay Mode and Gateway Switching Equipment for the Interconnection between China Unicom's GSM Network and the Public Main Networks" in June 1995. This document provided technical specifications pertaining to network interconnection between China Telecom and China Unicom. Although China Unicom had reservations about many of the terms of these specifications (affecting to find them "unfavorable"), the specification nevertheless was issued in the light of China Unicom's insistence for urgent network interconnection. Experience during the period from 1995 to 1998 has shown that these terms certainly placed China Unicom in a very vulnerable competitive position.

In general, the disadvantages conferred by these technical specifications on China Unicom can be summarized in the following sections [8, 10, 11].

A Long, Drawn-Out Process for Business Approvals

According to the master license issued by the State Council in 1994, China Unicom was immediately authorized to provide mobile phone service across the country. This meant that the company could legally provide service anywhere in China, without obtaining any special approval. According to the MPT, however, China Unicom was required to obtain approval separately for each geographical area. Documents that were submitted for approvals included information on network structure, capacity, and interconnection. As there was no time frame set for the approval process, China Unicom had to wait in a situation of protracted uncertainty. Normally, even if the MPT pronounced itself satisfied with the submitted documentation, it would take at least 60 extra days before China Unicom could obtain MPT approval. Reminiscent of the attenuation of the equipment approval process by British Telecommunications in the United Kingdom at the beginning of the 1980s, this long and unnecessary approval process artificially inflated network deployment time. Simultaneously, it also led to wastage in China Unicom's labor resource pool. The process required the establishment of a significant number of teams to deal with licensing issues in different areas. The secondary consequence of such a process was inflation of the cost base of the organization.

Restrictions over Mobile Switching Center Coverage

Traditionally, a local fixed network covers a geographically or administratively independent region, such as a city or a town. Due to the obvious mobility of mobile subscribers, the topology, control mode, and technology of a mobile network does not necessarily have to be the same as that of a local fixed network. In this case, mobile network coverage should be free from geographic or administrative restrictions, responding instead to the demand patterns of the market wherever they might arise. According to the technical specifications enunciated, however, each of China Unicom's mobile switching centers (MSCs) could only cover the same region as that of China Telecom's local fixed network, regardless of the size of the region and the capacity of the MSC. That is, the topology of China Unicom's mobile network must be the same as that of the China Telecom's fixed network. China Unicom thus had to install separate MSCs even in small cities and towns, although there were huge capacity surpluses in each of its MSCs. This dramatically increased network costs, involving, as was intended, duplication of plant and facilities. In contrast, China Telecom's mobile network was not subjected to this restriction. Its MSC could cover several regions that were covered by its fixed network, depending on the size of these regions.

Inefficient Cross-Network Relay Mode for Interconnection

According to the MPT's technical specifications, within the local network area in which China Unicom was providing mobile phone service, a dedicated gateway (GW) with switching functions had to be installed on both the China Unicom side and the China Telecom side for network interconnection. This relay mode is shown in Figure 4.2

According to Figure 4.2, the functions of GW_1 and GW_2 are simply for interconnection, and no switching function is necessary. Using two complex and costly exchanges as jumper terminals is undoubtedly a waste of investment. According to the MPT's technical specifications, however, GW must have a switching function, and both GW_1 and GW_2 should be installed and financed by China Unicom. This had led to a cost increase of nearly 5% to 10% for China Unicom, and the construction period was prolonged by at least 8 months.

Unfair Charging Scheme for Network Interconnection

In China, mobile subscribers are charged per minute, while fixed network subscribers are charged per unit (a unit being 3 minutes in length). According to the technical specifications, the interconnection charge between the mobile and fixed network was levied per unit. China Unicom's statistics, based on more than 20 million calls, revealed that the average calling time of its mobile subscribers was only 75 seconds, or 1.25 minutes. More than two-thirds of these calls lasted less than 1 minute. In this case, if the charging

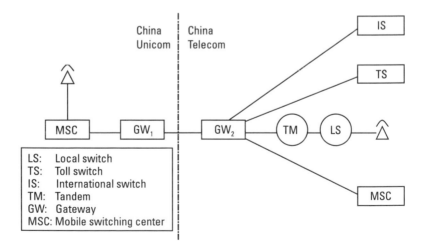

Figure 4.2 Cross-network relay mode for interconnection.

unit were changed into a single minute, the payment from China Unicom to China Telecom could be reduced dramatically.

Unfavorable Routing Plan

For calls between two networks, the technical specifications dictated that the long-distance call should be routed via the network interconnection point closest to the originating subscriber. Figure 4.3 shows the routing arrangement between China Unicom and China Telecom from 1995 to 1998.

As shown in Figure 4.3, when China Unicom's mobile user A originated a long-distance call to user B of China Telecom's fixed network, the call had to be routed via China Unicom's GW_1 to China Telecom's GW_1', and was then delivered over China Telecom's long-distance circuit. In this case, China Unicom had to pay China Telecom 92% of the charges it received from user A.

Indeed, to access mobile phone users, especially in the case of roaming service provision, it is necessary to access the mobile network via the gateway located at the originating subscriber's area. This is because the mobile system needs to clarify the location of its subscriber first, and then forward the call accordingly. In order to access fixed network subscribers, because the destination is definite, the call can be delivered directly to the destination via the

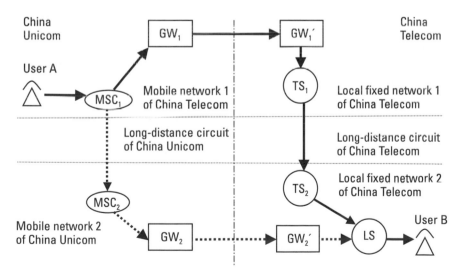

Figure 4.3 Routing arrangements for a long-distance call.

long-distance circuit of the originating party's network and then passed to the network of the other party at the destination. In fact, China Unicom has its own long-distance network specifically for long-distance traffic routing. If China Unicom uses its own internal long-distance circuit and accesses China Telecom's network via GW_2, then China Unicom could keep most of its long-distance income and pay China Telecom only 0.08 yuan for every 3 minutes. This, however, would militate against the terms outlined in the technical specifications.

In this case, it is obvious that the routing arrangement allowed for by the technical specifications was unfavorable to China Unicom, as all long-distance calls originating from China Unicom had to be delivered by China Telecom, no matter that they were terminated by China Telecom's fixed network or its mobile network. Most of China Unicom's income from long-distance service had to be passed on to China Telecom while China Unicom's long-distance network was simply being used to facilitate its roaming service.

The Signaling Problem

Because China Unicom was a new entrant, its mobile coverage was limited in its early years. For areas not covered by China Unicom, no gateways were installed. Therefore, if customer A of China Telecom's fixed network in these areas originated a call to customer B of China Unicom's mobile system in a different area, the call had to be forwarded by China Telecom's long-distance circuits to another area where China Unicom had installed a gateway. As is shown in Figure 4.4, however, the signaling system of China Telecom's network has made this process problematic in terms of processing billing information.

According to Figure 4.4, when customer A located in area A originated a call to customer B located in area B, the call had to be delivered to area C, by default, due to the fact that China Unicom had not yet established its mobile network in area A. In area C, the call was transferred to China Unicom via GW_C and $GW_C{'}$, and was then rerouted through China Unicom's proprietary long-distance circuits to area B, or to another area in the case of roaming. Certainly, this call should be defined as a long-distance call, as China Unicom's long-distance circuits were used. However, as the signaling signal from TS_A to TS_C was an initial address message (IAM), which was different from the initial address message with additional information (IAI) between LS and TS_A and did not carry the caller's identity code as the IAI does, China Unicom could not identify the caller's information and make relevant billing arrangements. As a result, a long-distance service could only

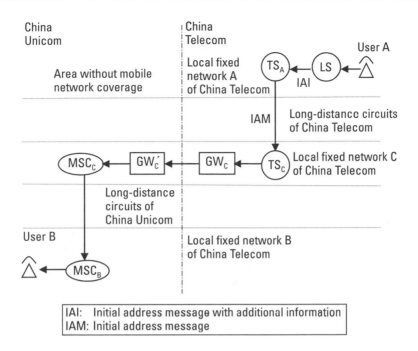

Figure 4.4 Call routing for areas not covered by China Unicom's mobile network.

be billed as local service due to the absence of complete information on the caller.

In fact, this problem could have been dealt with much more easily by upgrading the signaling system from IAM to IAI. China Telecom's signaling system was, in fact, configured for a situation of monopoly operation. Rationally, it should have been reconfigured and upgraded to accommodate the increasing openness of the telecommunications industry.

Quality Assurance of Network Interconnection

Once China Unicom had constructed its two gateways at each point of interconnection, the gateway on the China Telecom side was to be transferred to China Telecom for maintenance and operation. In this case, China Unicom was unable to monitor the operation of networks connected to the gateway in the China Telecom's side and thus could not make relevant routing arrangements. This created serious problems for quality of service. The MPT technical specifications did not contain any terms regarding quality of interconnection. Absent were defined standards on parameters, such as voice echo, line noise, cross talk, time delay, voice volume, and distortion. Many of

China Telecom's subscribers complained about the high congestion rate when calling China Unicom's mobile subscribers, while China Unicom simultaneously discovered that there existed a huge amount of spare capacity in the trunk connection between the two gateways. This meant that interconnection facilities were not being used effectively due to poor coordination between the gateway and the network at the China Telecom side.

Emergency Service

As the incumbent operator, China Telecom has been required to undertake certain universal service obligations, including free calls for emergency services, whereas China Unicom has not. According to the technical specifications, however, no network interconnection arrangement was set to exist between China Telecom and China Unicom for the provision of emergency services, such as 119 (fire brigade), 110 (police), 120 (ambulance), and 122 (traffic accident). This meant that China Unicom was required to install its own direct connections to individual emergency service providers and to bear the full cost.

The above issues are clearly suggestive. They are testimony to a close affiliation between the regulator and the incumbent operator during the period from 1995 to 1998 that left the new entrant in a vulnerable bargaining position in regard to network interconnection. Almost one-third of China Unicom's income was transferred to China Telecom during the period from 1995 to 1997, whereas China Telecom was required to pay back only 2% of this to China Unicom. The poor quality of interconnection damaged China Unicom's reputation. A new but inappropriately interconnected network is not dissimilar to an isolated network, and network externalities remain a concern for subscribers who would benefit from demand-side economies of scope. Given that China Unicom only occupied a minor market share in its initial stages, and interconnection quality was relatively poor, new subscribers tended to opt for the service provided by China Telecom. In so doing, quality communications to most other subscribers could be guaranteed. Telecommunications is an unforgiving sector. An "iron law" of the industry is that the utility of networks for the consumer is critically related to their size. The larger the network, the greater the utility and the greater the potential for further growth (positive feedback). Networks, naturally or artificially restricted in size, run the risk of diminution. The victim of a game in which its incumbent competitor was given across the board preferential status, China Unicom faced such a threat. By the end of 1997, China Unicom had only been able to acquire a risible 3% of mobile market share. As will be contended in the next chapter, this situation was destined to remain unchanged until 1998, when the telecommunications regulatory framework in China was restructured.

4.3   China Unicom: The New Entrant

While China Unicom suffered seriously from the imposition of a reversed asymmetric regulatory framework, the clear intention of the government remained: the deregulation of telecommunications. The powerful political and economic influence of its stakeholders and the growing atmosphere of liberalization, both nationally and internationally, conferred considerable strengths and advantages upon the organization.

The first such advantage emanated from China Unicom's entitlement to the benefits of favorable policies once bestowed upon the MPT by the Chinese government. These included a low tax impost, a high depreciation rate and the obligation to return only 10% of governmental loans [12]. The purpose of these policies was to give high priority to telecommunications development. Benefiting from similar policies, China Unicom was, theoretically, advantageously placed to achieve favorable operating conditions for its business compared with new companies in other industrial sectors. Viewed from this latter perspective, it had acquired impressive, if seemingly latent, comparative advantage.

The second advantage accruing to China Unicom lay in its flexible and diversified financial resources when compared to China Telecom. Although China Unicom could not be immediately listed in the stock market (as it was constituted as a state-owned company), it could readily raise the capital necessary to achieve its stated infrastructure goals by inviting state-owned institutions, including local government administrations, to be its financial stakeholders. This subjected China Telecom to considerable pressure. In the past, local government was the largest investor in telecommunications, clearly recognizing the importance of telecommunications as a key infrastructure for attracting inward investment and promoting the local economy. Traditionally, investments from local government were taken-for-granted by the MPT, seldom resulting in a financial return to their originators, while the government tended to concentrate on the macro benefits resulting from a well-developed telecommunications system rather than on issues of investment in telecommunications itself. The establishment of China Unicom, however, presented local governments with an attractive new option: They could still invest in China Telecom without any direct financial returns as they had done previously or, alternatively, they could divert investments to China Unicom and become its stakeholders, sharing the financial returns with China Unicom. An additional factor for consideration was that, by supporting the development of China Unicom in their region, the public could also gain benefits from competition. Obviously, the second option would be more attractive to local governments. The clear implication of this was that China Telecom would, to an important extent, lose local government support, both politically and financially. This constituted a formidable problem for China Telecom. Lacking limited company status, it was effectively precluded from issuing an invitation to local governments to become its stakeholders and consequently reap the potentially huge financial benefits.

A high-ranking official in the MPT revealingly offered two salient reasons why the MPT resisted turning its telecommunication operation over to a limited company in advance of market liberalization in 1994 (unattributed interview, October 18, 1994). The first lay in its reluctance to allow profits to be effectively diverted to other industrial sectors once their representatives became stakeholders in such a company. Second, the MPT could conceivably have been forced to relinquish its most favored company status (losing favorable government policy advantages) had it become a limited company. To obtain local governments' continuing support, China Telecom seriously considered handing over its rural telecommunications system to local government control. This proposal, however, was not well received in the local government sector. Low levels of profitability in the rural telecommunications market gave the impression that such control would endow local government with risks rather than benefits.

The third—critical—advantage accruing to China Unicom was political. The company's three sponsoring ministries (or champions) exerted key influence in the State Council stemming from their pivotal position as the sponsors of key industrial sectors in the Chinese economy. One director in China Unicom made clear to the authors that China Unicom "has three mouths in the Cabinet, while China Telecom has just one" (unattributed interview, October, 28 1994). Similarly, the 13 stock-holding corporations supporting China Unicom (referred to in Chapter 3) occupied a dominant economic position. China Unicom also benefited from close affiliation to the State Economic and Trade Commission, a key government department in the Chinese central administration. This again placed the organization in a very strong position of influence compared with China Telecom.

In addition to the cited example of intervention on network interconnection charges, another example of the powerful political potential of China Unicom centers on the allocation of frequency resources between China

Telecom and the company. The MPT has been extremely reluctant to allocate frequencies to China Unicom. After a protracted debate, Zou Jia-hua, a vice premier of the State Council, chaired a meeting attended by ministers from the MPT, the Ministries of the Electronic Industry, Railways, and Electrical Power. As a result of his intervention, important frequency allocations in the range of 800 to 900 MHz were partly allocated to China Unicom for its development of cellular phones and radio paging services.

Senior Chinese leaders have also conferred political legitimacy on China Unicom through public endorsement of the company. In his speech at the opening ceremony, Vice Premier Zou Jia-hua explicitly pointed out the importance of sponsoring China Unicom:

> The importance of supporting China Unicom should be recognized within the context of reform of the national economic system. Some comrades thought that support for China Unicom means support for some ministries holding China Unicom's shares. Essentially, the establishment of China Unicom is a high level decision made by the State Council. It is a strategy directly related to the national interest. Both President Jiang Ze-min and Premier Li Peng have participated in the process of policy making establishing the company.
>
> The establishment of China Unicom is an innovative public policy issue and a substantially new policy redirection. As a new company, the initial stage of China Unicom will be very difficult, as it has to face the challenge from the existing economic system, conservative mentalities and traditional management styles. If we do not support it positively, then it will be very difficult for China Unicom to grow. [13]

Such statements from central government are critical to the creation of a supportive environment for new entrants and particularly significant given the nuances of the Chinese political context.

The fourth advantage accruing to China Unicom lay in its efficiency potential. As a newly established company and new entrant, China Unicom has been in a position to adopt an organizational structure simpler in form than the bureaucratic structure that was such a central feature of China Telecom. It was able to adopt a flatter management structure and was not faced with the encumbrances of the incumbent provider. It was neither burdened with a heavily hierarchical structure nor carried the costs of a vastly inflated workforce. Furthermore, China Unicom was in a position to avoid the corporate pitfalls of the Chinese social security system. Pensions provision has traditionally been the responsibility of the organization within which the pensioner has been employed. Such a provision works to the disadvantage of the incumbent, which has, over the years, acquired a heavy actuarial burden. In 2002, the Chinese social security system is being reformed along the lines of the Western model. In this sense, no such handicap as yet impinges on China Unicom, and probably will not in the immediate future. This, if coupled with effective operational management, should equip the challenger with a clear competitive advantage. Heavy liabilities for pensions, for example, constitute a burden of stranded investment on the incumbent.

Finally, China Unicom has not been subjected to a universal service obligation. This has made it possible for the company to "cream skim" in areas of high demand while avoiding the necessity of providing obligatory service in high-cost rural areas. Since its establishment, the company has strategically targeted the profitable cellular phone market in the large and well-developed cities where demand is high (and costs relatively low) with the enticing prospect of amortizing investment within months rather than years.

These advantages have conferred great competitive potential and comparative advantage on China Unicom. As a new market entrant, however, the company has inevitably exhibited weaknesses and encountered competitive barriers. The first and most critical barrier has been network externality. Due to considerations arising from embedded long-term self-interest, the MOR and the MEP did not donate their private networks as a dowry to China Unicom. As a result, the company has been obliged to construct its own network, over a protracted period of time. Predictably few customers have been attracted to this small network due to the effect of network externality, particularly the fact that network connection with the PSTN has been poor. As a result, China Unicom has had to concentrate on mobile phone service because of the limited investment involved and the short period needed for rolling out the network. On July 19, 1995, the first anniversary of the establishment of China Unicom, four GSM mobile phone systems with a respective capacity of 100,000 subscribers were formally launched in Beijing, Shanghai, Tianjin, and Guangzhou. Vice Premier Zou Jia-hua attended the opening ceremony in Beijing and became its first subscriber [14].

The second barrier has been customer inertia with few customers willing to accept the inconvenience of changing carriers. The incumbent has benefited from formidable customer loyalty and habit since it was established as the sole telecommunications operator in 1949. Although many customers expressed dissatisfaction with the service quality and poor customer relations of China Telecom, they have shown a marked reluctance to fully embrace a new operator, particularly in such services as mobile phones where China Telecom was fully able to meet market demand (no waiting list) and had begun to improve its service as a result of the entry of China Unicom.

Although China Unicom's handsets were sold at a very competitive price, most customers have still been inclined to subscribe to China Telecom's network [15]. As a result, China Unicom has been forced to project its image hard and spend heavily on image building and product promotion.

The third barrier has arisen from the fact that China Unicom, as incomer, obviously lacked experience in the operation and management of a telecommunications system, as none of the three ministries sponsoring China Unicom had previously specialized in telecommunication services. Accordingly, China Unicom has had to quickly ascend a steep learning curve in order to master such skills in this dynamic and fast-moving sector.

A significant deficit in managerial and technological talent constituted the fourth problem facing China Unicom. In order to develop its services, China Unicom required a cadre of telecommunications professionals, armed with skills in engineering and management. Most of its employees, however, originated from three shareholding ministries, and typically lacked experience in telecommunications work. To expand its professional cadre of human resources, China Unicom needed to recruit newly graduated university students and attract in talent from the MPT and China Telecom—the most obvious "instant" source of trained supply. As explained below, however, the discrete characteristics of the Chinese educational and personnel management systems have rendered this very difficult.

In China, most universities were sponsored by and affiliated with relevant ministries before educational reform in 1998. The MPT, for instance, enjoyed vertical integration with five universities located in Beijing, Nanjing, Changchun, Xian, and Chongqing. There also existed at least one middle-level training institute of posts and telecommunications in each of the 30 provinces, which belonged to the MPT's provincial branches (PTAs) and was used mainly for training mid-level technicians. Every year, the bulk of graduates from this supply chain were recruited by the MPT's affiliated enterprises and research institutes in line with the MPT's human resource plans, and only a small fraction (usually 5%) were required by the State Education Commission to be available for recruitment by other industrial sectors. Under such circumstances, it was inevitable that China Unicom would be forced to recruit from other universities, targeting students majoring in the broader field of electronic engineering rather than specifically in telecommunications. This situation remained unchanged until other universities launched telecommunications programs in more recent years. Before sufficient numbers of students had obtained their degrees, China Unicom was forced to attract talent directly from China Telecom, which, as indicated below, has proven to be a process fraught with difficulty.

When China Unicom was established in 1994, the State Council requested the MPT to lend full support to the company by providing qualified talent in line with reasonable demands from China Unicom. To comply with this, the MPT provided a list of 100 named professionals to China Unicom. The latter professed dissatisfaction with the nominated candidates and refused to accept them, as the suspicion lurked that the MPT was unwilling to voluntarily make available its highest prized human assets. Undaunted, China Unicom continued with its attempt to mine that talent for itself from among the employees of China Telecom—and this in the face of the construction of a series of harassing impediments by China Telecom. For example, when China Unicom attempted to employ a chief engineer from a provincial PTA, the engineer and his family were told to move out immediately from the apartment provided by the PTA. Such activity on the part of the incumbent constituted a major threat to the fledgling China Unicom. As a new company lacking in fixed assets, such as a supply of suitable residential stock, it could not turn on the supply of accommodation at will. In China, the tradition has been that it is the employing units or companies that typically provided residential apartments. Although the Chinese government has been trying to reform the housing system since the early 1990s, this is understandably proving to be a protracted process. In this situation, China Unicom had to invest heavily in the construction of apartments to accommodate its newly recruited staff following its establishment.

Another serious threat to China Unicom has been the lack of a legal framework in the increasingly decentralized telecommunications market. To date there is a marked absence of a telecommunications law. The process of drafting such a law began in the late 1980s but has been paralyzed by squabbling among major ministries, chief among which were the MPT and the MEI. Several drafts of the telecommunications law drawn up by the MPT have been effectively squashed by the MEI and other ministries [14].

The absence of an effective telecommunications legal framework in China has made the achievement of effective regulation even more intractable, conflictual, and complicated [16]. Disputes between China Unicom and China Telecom have had to be remitted to the State Council, which has attempted to act as honest broker in the search for political resolution of this issue. Typically neither side has been fully satisfied with resulting settlements, as there have been no stable rules for the State Council to follow, a situation requiring disputes to be resolved on a case-by-case basis. The result has been that competition in the market has been paralleled by administrative competition as the various players in the telecommunications game seek comparative advantage under the aegis of State Council adjudicative decision making.

While clearly cognizant of its weaknesses as well as its strengths, China Unicom has quickly established a reputation as an aggressive and confident new entrant facing unique market opportunities and consequent political advantages. Since its establishment in 1994, the company has formulated an ambitious development strategy and adopted an aggressive stance to enable it to penetrate the market and seriously compete with the incumbent.

As part of this development program, China Unicom has fully utilized its situational advantages to compete effectively with China Telecom. The first evidence of this came in the signing of a long-term memorandum of understanding between China Unicom and the American General Telephone and Electronic Corporation (GTE) on January 15, 1995 [17]. Under the terms of this memorandum, the partners set up a joint venture to undertake design and construction projects for China Unicom. This was a strategically important method for bypassing the design and construction institutions of the MPT, which are effectively prevented from cooperating with China Unicom due to their close affiliation with the MPT. In addition, according to the memorandum, a technical research and development center would be jointly established, mainly focused on conducting research and developing new technologies and telecommunications products for China Unicom. The establishment of this center and the maturation of this joint venture have enabled China Unicom to access the most advanced telecommunications technology. Their American partner has effectively conferred upon the company the advantage of a strong technological second mover. For GTE, the advantages lie in establishing an opportunity for penetrating the potentially vast Chinese marketplace. Harvey Greisman, a GTE vice-president, announced that the U.S. company envisaged a "long-term strategic alliance with China Unicom" [18]. The significance of the memorandum may be gauged from the fact that cosignatories to it were Zou Jia-hua, the vice premier, and Dr. Henry Kissinger, the former Secretary of State of the United States. The signatures of the head of the State Economic and Trade Commission and ministers from the three ministries who were cofounders of China Unicom were also on the agreement. Such backing again revealed the growing political influence clearly enjoyed by China Unicom.

It was announced at the same time that Dr. Henry Kissinger was invited to be the honorary adviser to Mr. Zhao Wei-chen, then General Director of China Unicom. This was an important public relations coup for the company given Dr. Kissinger's influence in China beginning with his celebrated diplomatic visit in 1972, which had resulted in the establishment of a path-breaking working relationship between the People's Republic and the United States after more than 20 years of mutual hostility.

China Telecom's reactions to China Unicom's moves were not difficult to predict. The incumbent had long exhibited an "unwillingness to share the lucrative domestic telecommunications market" and endorsed the orthodox view of some State Council members amounting to a policy of "monopoly in one country" [19]. The agreement between GTE and China Unicom offered a direct challenge to such inward looking thinking and one of deep symbolic political significance. It suggested that, in its short history to date, China Unicom had successfully penetrated the inside track of public policy making in this strategic sector of the Chinese economy.

At the same time, very aggressive marketing strategies adopted for competition with China Telecom on mobile phone service—the sole service provided by China Unicom before 1998—have also been much in evidence. For instance, prior to the establishment of China Unicom, subscribers could be connected to MPT's mobile phone networks only after they bought a handset from the MPT. In other words, the MPT was not only a monopoly mobile service provider, but also the sole retailer of mobile handsets. This monopoly status of the MPT made it possible for the ministry to make a high profit through market gouging simply by tied selling of handsets at an unreasonably high price. In 1992, the price of a handset plus connection fee stood as high as 30,000 yuan ($3,636). The policy adopted by China Unicom, however, was that customers could use any handset, no matter from whence it was sourced, to connect with its mobile phone networks. As a result, the monopoly in the handset market was effectively broken and handsets were immediately made available in numerous retail outlets. In consequence, the price of each handset fell dramatically. Figure 4.5 shows the impressive decline in handset prices and connection fees since China Unicom entered the market in 1994.

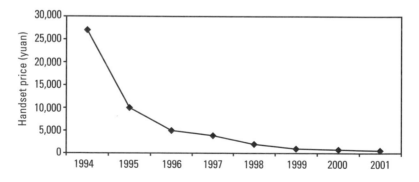

Figure 4.5 Handset and connection price reductions. (*Source:* China Unicom.)

Another bold experiment undertaken by China Unicom lay in the so-called China-China-Foreign (CCF) investment model [20]. To turn China Unicom's ambitious development blueprint from dream to reality, it was estimated that $15 billion would be required before 2000. It would be extremely difficult to raise such investment capital solely from the country's domestic capital market. However, foreign direct investment in Chinese telecommunications had long been banned, and no foreign corporations were allowed to acquire ownership or operational rights in Chinese public telecommunications before China's accession to the WTO in 2001. Under these circumstances, innovative company strategies were required to bypass restrictive government policies.

According to the CCF model, a company owned by local government or one of China Unicom's stakeholders could form a joint venture with foreign investors. Under this arrangement, such a joint venture would build up the network while China Unicom was responsible for the network operation. Cash flow, such as income from installation fees and service charges, would then be shared between China Unicom and the joint venture (Figure 4.6). In this way, China Unicom has managed to attract 70% of its total investment from foreign investors [20]. By April 1998, 23 projects with a total investment of $1.5 billion had been invested through the CCF scheme [21]. GTE was one of the first group partners in China Unicom's CCF scheme. GTE's efforts in the Chinese telecommunications sector bore early fruit.

The above strategies adopted by China Unicom indicate that the company has clearly scanned its environment, both internal and external, and has

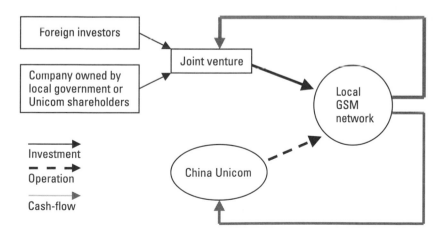

Figure 4.6 China Unicom's CCF investment scheme.

adopted an increasingly aggressive competitive stance towards China Telecom by fully utilizing the advantages it has enjoyed. It has indicated to China Telecom that it is fully prepared to engage in a head-to-head competitive game. In this sense, it is not extravagant to assert that the process of telecommunications deregulation has been well under way in China since the mid-1990s [22]. It is apparent from foreign experience, however, that the full benefits of such a competitive situation are unlikely to be readily achieved in the absence of a developed regulatory framework. To obtain the results of competition fully, the reversed asymmetric regulatory framework—a marked feature of the Chinese telecommunications sector—had to be effectively restructured.

4.4   Conclusion

In summary, the Chinese telecommunications policy experiment during the period 1994 to 1998 is best to be seen as part of an unfinished revolution in this locomotive sector of the Chinese economy. This revolution continues to unfold. In conceiving and supporting the establishment of China Unicom, the Chinese government has signaled its recognition of the significance of telecommunications and its key role in the development of the new economy. However, the essential ambiguity of the notion of "one nation, two systems" (itself a recognition of the functional necessity of adopting a form of pseudo-capitalism in a situation of social transformation) was revealed in the operational handicapping of China Unicom. An incomplete and essentially reversed asymmetric system of regulation in which the new market entrant rather than the incumbent has been disadvantaged by regulatory constraints has been revealed in a series of entry or operational barriers imposed on the new entrant; these have required its strategic decision-making coalition to display patience, fortitude, and innovatory zeal in its efforts to compete against the incumbent. Reminiscent of the character building lauded by such ancient writers as Sun Tze as required for success in conventional (or commercial) war, such Chinese characteristics appear to have stood China Unicom in good stead in its efforts to benefit from and play a key creative role in the development of competition within a liberalized telecommunications environment.

Without doubt, the entry of China Unicom may be presented as the arrival of the fox in the chicken coop. Its proactivity constituted a major threat to the dominance of China Telecom, which has characteristically sought to manage the threat of the precocious new arrival by erecting, in orthodox fashion, obstacles to strew in China Unicom's path. Despite such misanthropic endeavors, however, China Unicom has weathered such

problems and demonstrated that it is a telecommunications player "with attitude." Far from recoiling in the face of such resistance, the organization has
deployed considerable political skill, resilience, and commercial acumen in
its efforts not only to effect market entry but also achieve long-term sustain-
ability. Such efforts have been marked by a campaign of recruitment leading
to effective human resource development in the service of a "close to the customer" approach. Lacking the bureaucratic inertia and legacy of sunk investments typical of a large incumbent organization, China Unicom has been
able to achieve a high degree of organizational leanness allowing it to react
quickly to market trends. Its attempts to woo impressive foreign partners,
such as GTE, reveal an organization more than willing to break out of the
traditional corset of Chinese autarkic thinking.

In all this, of course, China Unicom has hardly acted as a political iso-
late. On the contrary, its cleverly crafted policy of obtaining support in the
State Council and the political leverage emanating from its sponsoring Ministries have conferred upon the organization an image of political astuteness
contrasting markedly with the somewhat desperate efforts of China Telecom
to constrain its new rival.

Clearly, it is premature to conclude that the Chinese telecommunications "revolution" has run its full course. Succeeding chapters will point to
the still unfolding nature of that revolution. It is not extravagant, however, to
conclude that the period of policy experimentation marked by the establishment of China Unicom is of huge significance in the economic biography of
twentieth and twenty-first-century China. In this regard China Unicom is to
be seen as a bellwether organization, which, like the sentinel species in biol-
ogy, can be seen as a portent of wider environmental transformation.

References

[1]   Ministry of the Posts and Telecommunications, *China Posts and Telecommunications Brochure*, 1994.

[2]   People's Posts and Telecommunications Newsletter, March 3, 1994.

[3]   China Telecom Newsletter, July 1995.

[4]   Gao, P., and K. Lyytinen, "Transformation of China's Telecommunications Sector: A Macro Perspective," *Telecommunications Policy*, Vol. 24, Nos. 8–9, 2000, pp. 719–730.

[5]   Mesher, G., and E. E. Zajac, "Toward a Theory of the Global Liberalization of Tele-communications," paper delivered at 12th Biennial Conference of the International Telecommunications Society, Stockholm, Sweden, June 1998.

[6] Noll, R. G., and F. M. Rosenbluth, "Telecommunications Policy: Structure, Process, Outcomes," in *Structure and Policy in Japan and the United States*, P. Cowhey and M. McCubbins (eds.), Cambridge, UK: Cambridge University Press, 1995, pp. 119–176.

[7] Xu, Y., and D. C. Pitt, "One Country, Two Systems—Contrasting Approaches to Telecommunications Deregulation in Hong Kong and China," *Telecommunications Policy*, Vol. 23, Nos. 3, 4, 1999, pp. 245–260.

[8] Xu, Y., "The Impact of the Regulatory Framework on Fixed-Mobile Interconnection Settlements: The Case of China and Hong Kong," *Telecommunications Policy*, Vol. 25, No. 7, 2001, pp. 515–532.

[9] State Planning Committee Document, No. 14101.

[10] He, G. P., "Various Technical Problems in the Interconnection Between China Unicom's GSM Network and P&TS PSTN Network," *China Communications*, July 1998, pp. 32–35.

[11] Xu, Y., D. C. Pitt, and N. Levine, "Interconnection: A Bottleneck to Future Chinese Telecommunications Deregulation?" in *21st Century Communications Networks*, P. Enslow, P. Desrochers, and I. Bonifacio (eds.), Amsterdam, the Netherlands: IOS Press, 1997, pp. 106–114.

[12] Chen, Y. Q., "Driving Forces Behind China's Explosive Telecommunications Growth—Changes in Policies Fuels Growth," *IEEE Communications Magazine*, Vol. 31, No. 7, 1993, pp. 20–22.

[13] Zou, J. H., "Speech at the Opening Ceremony of China Unicom," *People's Posts and Telecommunications Newsletter*, July 21, 1994.

[14] *People's Daily*, July 20, 1995.

[15] China Telecom Newsletter, May 1995.

[16] Pitt, D.C., N. Levine, and Y. Xu, "'Touching Stones to Cross the River'—Evolving Telecommunications Policy Priorities in Contemporary China," *J. Contemporary China*, Vol. 13, No. 5, 1996, pp. 347–365.

[17] *People's Daily*, January 16, 1995.

[18] *Financial Times*, January 17, 1995.

[19] Tan, Z. A., "Challenges to the MPT's Monopoly," *Telecommunications Policy*, Vol. 18, No. 3, 1994, pp. 174–178.

[20] Li, H. F., "Speech at the News Conference Celebrating the Third Anniversary of China Unicom," July 19, 1997.

[21] *Ming Pao Daily*, April 23, 1998.

[22] Xu, Y., and D. C. Pitt, "Competition in the Chinese Cellular Market: Promise and Problematic," in *The Future of the Telecommunications Industry—Forecasting and Demand Analysis*, D. G. Loomis and L. D. Taylor (eds), Boston, MA: Kluwer Academic Publishers, 1999, pp. 249–264.

5

New Horizons? Restructuring the Regulatory Framework

To safeguard the competitiveness of telecommunications markets, "it is essential to establish an impartial judge to frame the rules and regulations for the orderly development of the sector and to adopt an appropriate mechanism by which the whole telecommunications industry is to be regulated" [1]. For this reason, the Basic Telecommunications Agreement of the WTO suggests that member countries should be committed to establishing an independent regulatory body that "is separate from, and not accountable to, any supplier of basic telecommunications services. The decisions of and the procedures used by regulators shall be impartial with respect to all market participants" [2]. Strictly speaking, an independent regulatory agency should also remain at arm's length from the government, because when a regulatory body is simultaneously a government department, it is normally vulnerable to political intervention [3].

Experience in many countries, however, has shown that transition from a traditional regulatory framework, with government responsible for both operations and regulation, to a single independent regulatory agency can be a slow and contentious process. Many late-mover countries have learned this lesson the hard way [4]. In China, the controversial settlement over network interconnection (as reviewed in Chapter 4) clearly indicates that an ineffective regulatory framework placed the new entrant in an unfavorable position in terms of competing with the incumbent operator. Consequently, the

91

government's ambition of achieving a competitive telecommunications market was seriously compromised.

5.1 The Context of Regulatory Framework Restructuring

To coordinate all the interested parties affected by the absence of a well-defined regulatory framework, China's State Council established a State Joint Conference on National Economic Informationization in 1994, which was later replaced by the State Council Steering Committee of National Information Infrastructure (SCSCNII) in 1996. It was chaired by one of the vice premiers, with participation by ministers from the MPT and China Unicom's shareholding ministries. Because of its weak legislative status, poor financial resources and lack of administrative power, however, SCSCNII failed to play an effective coordinative role [5, 6]. In principle, it acts as a forum in which all interested parties can argue and defend their self-interest, rather than as an agency to effectively and cohesively choreograph the competition. In the event, the SCSCNII's operation proved to be extremely inefficient. For example, it was committed to drafting a national informationization development strategy, but nothing emerged before the SCSCNII was dissolved in 1998 [7].

The ineffectiveness of the State Joint Conference on National Economic Informationization and the State Council Steering Committee of National Information Infrastructure in coordinating and balancing the interests of concerned parties has continuously placed China Unicom in a vulnerable position when competing with China Telecom. China Unicom and its shareholding ministries have since made strong appeals for restructuring of the regulatory framework and, in particular, complete functional and organizational separation between China Telecom and the MPT.

In addition to pressure from China Unicom and its shareholding companies, the rapid expansion of private networks, particularly the broadcasting network of the Ministry of Radio, Film, and Television, has also impelled the central government to establish an effective regulatory framework to coordinate different sectors and prevent duplicative network construction. In fact, almost all projects constructed by non-MPT sectors were to be categorized as duplicative construction as long as China Telecom possessed similar systems. Some senior government officials even considered China Unicom's network as a prime example of such duplicative investment. This argument formed the crux of the former MPT's rationale for treating China Unicom unfavorably.

One of the main reasons behind the rapid expansion of private networks in China since the mid 1990s has been the extremely high rental fee and notoriously bad service quality provided by China Telecom. For example, it cost the Shanghai Radio, Film, and Television Administration only 20 million yuan to build a private optical network, while it would have cost the former company 54 million yuan to lease analogous network resources from China Telecom over a period of 3 years with the same area coverage [8]. This kind of economically unsound deal and discriminatory pricing tactics have been the main rationale for the Ministry of Radio, Film, and Television and other ministries to lease public networks only with the greatest reluctance. In consequence, private networks have developed rapidly. Such developments have been fuelled by realization of the massive market potential of telecommunications.

These private networks have raised considerable concerns on the part of the government, as they are perceived as threatening with redundancy the public network in which the government has invested heavily in the past 20 years. Additionally, since all these private networks are, in actual fact, owned by the state, the budget-strapped government has viewed such duplicative investments as a considerable waste of public resources. In consequence, preventing duplication and improving the utilization of the existing public network became one of the major incentives for the government to restructure the regulatory framework. In his report to the People's Congress, Luo Gan, the General Secretary of the State Council, stated explicitly that the integration of telecommunications, broadcasting, and computing networks should be a major principle behind any reform of the regulatory frameworks of the information industry [9].

In March 1998, the Ninth National People's Congress, the supreme legislative body of China, approved an institutional restructuring scheme proposed by the State Council. According to this scheme, the administrative system of the Chinese information industry was to be subject to drastic revision. On April 1, 1998, the MII was officially established as a result of amalgamation of the former Ministry of Posts and Telecommunications and the Ministry of Electronic Industry.

According to the State Council, the major commitments of the newly established MII are as follows:

- To revitalize the electronic and information technology manufacturing sector, the telecommunications sector, and the software sector;
- To promote the informationization of the national economy and social services;

- To make plans, policies, and regulations for the information industry;

- To conduct overall planning and regulation for the national backbone communications networks (including both local and long-distance telecommunication networks), the radio and television broadcasting networks (including both cable and non–cable TV networks), as well as the private networks owned by the military and other sectors;

- To effectively allocate resources to avoid duplicative construction and safeguard information security.

The State Council indicated that regulation of the broadcasting network, once the responsibility of the former Ministry of Radio, Film, and Television (MRFT), and the private telecommunication networks of the Aerospace Industry Corporation and the Aviation Industry Corporation, would be transferred to the MII. The MII would also take control of the newly established State Postal Bureau. Simultaneously, the State Council Steering Committee of the National Information Infrastructure, a nonstanding governmental body, would be dissolved, while the State Radio Regulatory Commission would be incorporated into the MII as the Radio Regulatory Department.

The transference of regulatory power over broadcasting networks came to a premature end because the former MRFT, since reconstituted as the State Administration of Radio, Film, and Television (SARFT), was reluctant to hand over its power. The official excuse proffered by the SARFT was that broadcasting constituted the "tongue and throat" of the Chinese Communist Party, and therefore should be administered separately from the public telecommunications networks. It lobbied the Propaganda Department of the Central Committee of the Communist Party to intervene and effectively frustrate such transference of power. In fact, the true reason behind this tactic was the SARFT's clear realization of the commercial value of the cable network in the rapidly growing telecommunications industry. The conflict between China Telecom's local network and the cable network of SARFT in recent years, as will be discussed in Chapter 7, clearly demonstrates the SARFT's strategic ambition to move into the telecommunications market. As a consequence of the SARFT's objection, the MII failed to attain status as a regulator able to coordinate a spectrum of industrial sectors that continue to move increasingly towards convergence.

Nevertheless, restructuring of the regulatory framework in 1998 was the most drastic institutional reform ever undertaken in the Chinese telecommunications industry. If the establishment of China Unicom in 1994 symbolized the first wave of liberalization in the Chinese telecommunications market, the creation of the new regulator—the MII—could justifiably be claimed to be a milestone in the second round of telecommunications deregulation in this late-mover country.

5.2 The New Regulatory Framework

Importantly, this institutional restructuring enabled the reduction and rationalization of the numbers of government departments involved in the information industry sector. As a result, administrative power over regulation became more concentrated. Table 5.1 summarizes the change that has happened in the restructuring of the regulatory framework.

A new regulatory framework for the information industry was formed in the wake of institutional restructuring. The main characteristics of the resulting framework, in terms of power allocation and operational structures, are as follows [10]:

- The MII was designated by the State Council as the regulator of the national information industry. With the powers granted and defined by the State Council, the MII is committed to drafting and implementing sectoral plans, policies, and regulations for the entire information industry sector. This new arrangement embraces the entire arsenal of power once possessed by the former MPT and MEI. Additionally, the regulator's power was also extended to the regulation of private networks, once owned and controlled by individual ministries, although the transference of power over cable network regulation came to a premature end. The MII is also committed to the overall planning of all types of public and private telecommunication networks and the effective allocation of resources while acting as the guarantor of information security. In addition to the above, the MII was also destined to take over part of the power once possessed by the former State Council Steering Committee of the National Information Infrastructure and State Radio Regulatory Commission. As a result of this transference of power and influence, the MII was placed in a predominant position of influence in the overall administrative and regulatory framework of the national information industry.

Table 5.1
Comparison of Changes in Information-Industry-Related
Governmental Departments
Before and After Restructuring

Before Restructuring	After Restructuring
I. Cross-Ministry Liaison and Coordination Body	
State Council Steering Committee of National Information Infrastructure	Dissolved
II. Cross-Ministry Regulatory Body	
State Radio Regulatory Commission	Merged with the MII
III. State Council Macrolevel "Fine-Tuning" Departments	
State Planning Commission	State Development Planning Commission
State Economic and Trade Commission	State Economic and Trade Commission
Ministry of Finance	Ministry of Finance
People's Bank of China	People's Bank of China
The State Economic System Restructuring Commission	(became a top-level think tank)
IV. Ministries of the State Council for Individual Industrial Sectors	
Ministry of Posts and Telecommunications	Merged with MII
Ministry of Electronic Industry	Merged with MII
Ministry of Radio, Film, and Television	State Administration of Radio, Film, and Television
Ministry of Foreign Trade and Economic Cooperation	Ministry of Foreign Trade and Economic Cooperation
V. State Council Science, Education, Culture, Society, and Resource Administrative Departments	
State Commission of Science and Technology	Ministry of Science and Technology
State Commission of Education	Ministry of Education

- The State Development Planning Commission, the State Economic and Trade Commission, the Ministry of Finance, and the People's Bank of China (the sole central bank of China) constitute the macroeconomic fine-tuning departments of the Chinese government. These departments are responsible, respectively, for the allocation of national resources, financial management, and banking administration. Such fine-tuning is conducted in accordance with state development strategy, the government's guiding principles, macroeconomic objectives, and the special circumstances of each specific industrial sector. The strengthening of the strategic role of the information industry in the overall national economy and the provision of favorable policy support to the information industry are both actuated through these departments. In decision making on key issues, the MII is obliged to cooperate with all of these departments. For instance, the development plans of the information industry must be rendered consistent with the comprehensive goals of the State Development Planning Commission, while a favorable loan policy for the information industry must be approved by the People's Bank of China. Following interdepartmental negotiations, final decisions on key issues reside with the State Council. The overall importance of the State Council was highlighted by the fact that it was this body rather than the MII that defined the terms of China's commitment to the WTO in the telecommunications sector.

- The Ministry of Foreign Trade and Economic Cooperation, the Ministry of Science and Technology, and the Ministry of Education constitute the key administrative departments responsible for foreign trade, economic cooperation, science and technology advancement, and human resources training and development. These departments are able to provide support for and exert their influence on state informationization policy and strategy. A close cooperative relationship with these departments is critical for the MII. For example, all of the major joint venture projects of the information industry are required to obtain approvals from the Ministry of Foreign Trade and Economic Cooperation, while the importation and upgrading of information technology require support from the Ministry of Science and Technology. Additionally, according to the State Council, all universities that were previously affiliated with the former MPT have now reverted to the Ministry of Education. The difficulty once encountered by China Unicom in sourcing new university graduates from these universities no longer exists.

- Private networks that were previously owned by the Ministries of Railways and Transportation, the Bureaus of the Coal Industry, Petroleum, Forestry, and other subsidiary departments under the State Council and the State Economic and Trade Commission are all subject to MII regulation. The same situation pertains for private networks owned by electrical companies, banks, and other commercial sectors. The general development of private networks is subject to the administrative guidance of the MII, while major private network projects must be assessed for approval by the MII. Such networks, however, will continue to be owned and operated by their original owners.

- Provincial governments are committed to the promotion of informationization in their provinces. Each provincial government has established a Department of the Information Industry. A selection of officials of the former provincial PTAs have joined these departments, while the rest have remained in the MII's newly streamlined branch: the Provincial Communications Administration. The function of this latter administration is mainly to regulate the provincial telecommunications market in accordance with regulations formulated by the MII.

In addition to the above government departments that are included in the main organizational constellation of the information industry, others can also exert their influence on sectoral developments. Such departments have responsibility in the following areas: tax collection, price regulation, technology supervision, industry and commerce regulation, auditing, public security, legislation, state security, copyright, and intellectual property.

An array of highly specialized organizations complements the above in the information sector. One of these is the China Electronic Information System Promotion and Applications Office, which aims to facilitate the application of electronic information systems in different economic sectors. Another is the so-called Leading Group for the Application and Coordination of CAD, which is responsible for promoting computer aided design (CAD) in the industry. In addition, a group of nongovernmental organizations and associations provide assistance to governmental departments and serve as channels for the enhancement of communications among all sectors. Institutional restructuring had no direct impact on these organizations.

5.3   MII: The New Regulator

In the wake of organizational restructuring, and in addition to Provincial Communications Administrations in each province, there are 13 departments that constitute the MII (Figure 5.1). These are the General Office, the Department of Policy and Regulation, the Departments of General Planning, Science and Technology, Economic Reform and Economic Operation, Financial Adjustment and Clearing, Electronic Information Product Administration, Informationization Promotion (State Informationization Office), Radio Regulatory Department (State Radio Office), Foreign Affairs, and

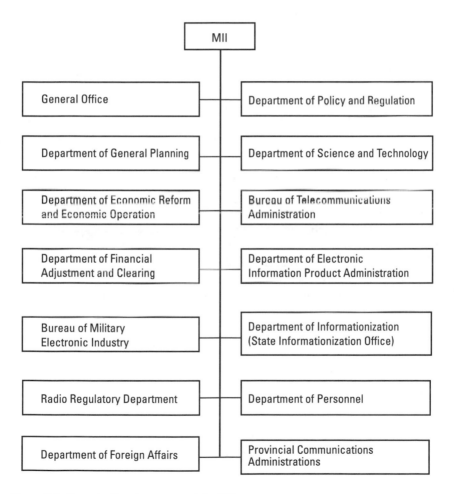

Figure 5.1 Organizational structure of the MII.

Personnel, the Bureau of Telecommunications Administration and the Military Electronic Industry, and the Radio Regulatory Department (State Radio Commission). Each of these departments and bureaus is charged with discrete responsibilities as outlined in Appendix A [11].

The establishment of the MII was undoubtedly a positive step towards further liberalization of the Chinese telecommunications market. The MII remains as a governmental department and not an independent regulatory agency in a real sense and thus lacks the level of independence from government enjoyed by the FCC in the United States and Oftel in the United Kingdom. The operational functions of China Telecom, however, have nevertheless been separated from the MII's regulatory commitments. According to the State Council, China Telecom is defined as one of the foremost large-scale state-owned enterprises in China and is under the direct supervision of the Central Enterprise Working Commission (CEWC), a newly established department of the Central Committee of the Chinese Communist Party. The CEWC is specifically responsible for the supervision of large-scale state-owned enterprises in recognition of the strategic importance of these companies in the Chinese socialist market economy, but it does not directly intervene in each individual company's routine operation. General managers of these key enterprises are directly nominated by the CEWC and endorsed by Wu Bang-guo, the vice premier of the State Council. In this circumstance, the MII currently enjoys a relatively neutral status in relation to telecommunications regulation because it is now independent from all operators and no longer shares any common interest with any operators. This status has enabled the MII to take a more procompetitive stance in facilitating competition in the Chinese telecommunication market by creating a more level playing field.

Soon after the establishment of the MII in March 1998, considerable progress was made in facilitating competition between China Telecom and China Unicom, including regulatory adjustment for network interconnection. The first major step lay in permitting China Unicom to provide services in any region without gaining separate permission from the regulator as long as these services were defined in its master license. This has greatly reduced the time and resources needed for China Unicom to launch its services in each individual region. Another favorable change resulted in China Unicom being permitted to use one MSC to cover more than one geographically or administratively independent region. This has reduced the overall cost for China Unicom and enhanced its network efficiency. The most recent regulatory move has obliged the incumbent to provide roaming service to China Unicom's subscribers in areas that have not yet been covered by China

Unicom's mobile network. Thus, China Unicom's subscribers can benefit from nationwide roaming services, as can subscribers of the incumbent. This should greatly reduce concerns about the network's geographic externalities. The MII also allocated new number blocks, such as 191, to China Unicom for its mobile service expansion [12].

5.4 Network Interconnection: A Reshaped Landscape

Given that China Unicom had expanded its service beyond mobile communications and adopted a more aggressive stance in entering fixed local, long-distance, and international telecommunication service markets, and that China Netcom and Jitong Communications Company have been issued licenses to provide telecommunication services [13], clear guidelines for network interconnection were urgently required. On September 7, 1999, the MII formally issued its "Provisional Regulation on Telecommunications Network Interconnection," the main contents of which are summarized below.

Obligations of the Dominant Operator

According to the regulation, an operator with more than a 50% share of the market is defined as a dominant operator. Such an operator is obliged to provide interconnection to all other operators requesting network interconnection at any technically possible and economically reasonable point, as long as network security is not compromised. The dominant operator is also obliged to provide all necessary network information concerning network interconnection to the party requesting interconnection, and is obliged to make relevant network modifications to facilitate interconnection.

As regards quality of service, the regulation clearly specifies that the quality of communications across networks should be the same as the quality of internal communications of individual networks. The dominant operator is also requested to provide interconnection within a defined time frame. Delays without a valid reason are not permitted.

Moreover, the dominant operator is obliged to provide interconnection for all services that it is currently providing to its own subscribers, including emergency services. The dominant operator must also provide directory enquiry services for subscribers of other operators. After negotiating the terms with other network operators, all subscribers should be able to access the other operator's number system via the dominant operator's enquiry system.

Technical Guidelines

The technical guidelines clearly define the points of interconnection and their locations. The gateway does not necessarily have to be an exchange dedicated to interconnection, and can be shared with other network facilities. It can also be shared between interconnecting and interconnected parties. This greatly reduces the costs incurred by China Unicom. Furthermore, subscribers have a choice of operators for their long-distance service, either on a subscription basis or on a call-by-call basis. Phone numbers are centrally allocated and administrated by the MII.

Cost Allocation for Interconnection

The new regulation prescribes that the party requesting the interconnection is not responsible for its full cost. Instead, the regulation specifies that each party must cover the cost of interconnection only on its own side, while retaining full ownership of interconnection facilities. The implication of this is that China Unicom is now only responsible for installing a gateway on its own side, matched by a commitment by China Telecom to install a similar gateway on its side. The party requesting interconnection, however, is obliged to cover the cost of installing trunk lines between the two gateways. The requesting party must also pay for renting ducts owned by another party.

In regard to interconnection charges, the regulation makes clear that these should be based on actual cost. Each party is expected to submit cost data to the MII, which, in turn, configures a settlement based on these costs with the help of an independent auditing agency. Until relevant cost data is available, interconnection charges will be based on the current retail tariff.

The regulation also defines other terms of interconnection, including the time frame for interconnection installation, the content of interconnection agreements, arbitration procedures, and penalty rules. This provisional regulation provides relatively clear guidelines on network interconnection. Although it is not as sophisticated as those in early-mover countries (for instance, there are no terms requiring accounting separation, a factor that could seriously influence the accuracy of the cost data), it is still an improvement on the situation in evidence between 1994 and 1997. In accordance with these guidelines, the MII revised the former MPT's 1995 technical specifications for network interconnection between China Unicom's mobile network and China Telecom's PSTN network. As summarized in Table 5.2, these revised terms have gone a long way towards addressing China Unicom's vulnerable position since 1998.

Table 5.2

Contrasting Regulatory Settlements over Fixed/Mobile Network Interconnection Before and After
the Regulatory Restructuring in 1998

	Before 1998	Since 1998
1	Service Provision Approval Procedure	
	China Unicom was required to obtain permission for service provision in each individual region, which normally took at least 60 days once all documents were ready.	China Unicom can provide services in any region under the terms of its master license without requesting separate permission for each individual region.
2	International Service	
	China Unicom's entire international traffic had to be routed through China Telecom's gateway and all revenues from international services were to be handed over to China Telecom.	China Unicom has obtained permission to construct its own international gateway and is able to retain all its international service revenues.
3	Interconnection Accounting	
	Interconnection charges were based on the old tariff, which did not reflect the real cost, given the outstanding monopoly operation of China Telecom.	Interconnection charges are to be based on cost once audited cost data is available.
4	Interconnection Cost Allocation	
	China Unicom should bear the full cost of interconnection including the installation of gateways on both the China Unicom and the China Telecom side.	China Unicom is only responsible for the installation cost of gateways installed on its own side plus the cost of trunk lines connecting the two gateways.
5	Gateway Functions	
	Gateways must possess independent switching functions and be installed separately from China Unicom's mobile system.	The gateway's switching functions can be shared by other network facilities and can be shared by both parties involved in interconnection.
6	MSC Coverage	
	Each MSC should cover only one geographically or administratively independent region, such as local fixed network, regardless of MSC's capacity and the size of the region—namely, the mobile network should have the same network topology as the local fixed network.	MSCs can cover more than one geographically or administratively independent region depending on the capacity of the MSC and the size of the region—namely, the mobile network can have different network topology from the local fixed network.

Table 5.2 (continued)

	Before 1998	Since 1998
7	Interconnection Charges	
	Interconnection charges for mobile to fixed calls was 0.08 yuan per unit (3 minutes), while the average calling time was 1.25 minutes per call.	The interconnection charge has been reduced to 0.05 yuan per unit but the unit remains 3 minutes in length.
8	Long-Distance Call Routing	
	All of China Unicom's long-distance calls were routed via China Telecom's long-distance circuits; China Unicom had to forward 92% of its long-distance service revenues to China Telecom.	Subscribers have a choice as to which operator to use for their long-distance calls, either on a subscription basis or on a call-by-call basis; however, the 92% reallocation rate remains valid.
9	Signaling System	
	The original signaling system used by China Telecom was configured for monopoly operation and China Unicom could not receive the callers' numbers for long-distance calls; hence, it could not issue billing invoices properly. The old Signaling System 1 was sometimes used for interconnection.	Signaling System 7 has to be used for network interconnection, and gateways on each side should have detailed billing capacity.
10	Emergency Services	
	China Unicom had to install direct connections to each individual emergency service provider, while China Telecom was obliged to provide interconnection for basic phone services only.	China Telecom is now obliged to provide interconnection for all services it is currently providing for its own subscribers, including emergency services.
11	Quality of Interconnection	
	No quality indicators were defined for network interconnection and no deadline was set for implementing interconnection.	A time frame for completing interconnection is imposed: 7 months from time of request. Quality indicators are as follows:
		1. The transmission quality for internetwork communications and intranetwork communications should be the same.
		2. The call failure rate between the MSC and the toll switch (TS) should be less than 1%, and the average traffic between them should be less than 0.70E in peak time.
		3. The call failure rate between the MSC and the tandem (TM) should be less than 0.5%, and the average traffic between them should be less than 0.70E in peak time.

Table 5.2 (continued)

12	Network Externalities	
	For areas not covered by China Unicom's mobile network, subscribers could not benefit from a roaming service.	The incumbent is obliged to provide roaming services for China Unicom's subscribers in areas that have not yet been covered by China Unicom's mobile network.

Source: MII, China Telecom, China Unicom, China Communications Info (http://www.cci.cn.net).

As a result of the improvements made to the existing network interconnection regime and other regulatory arrangements by the newly established regulator, China Unicom has achieved rapid network expansion in recent years. Figure 5.2 shows the evolution of China Unicom's market share since it formally launched its mobile service in 1995, which has jumped from less than 6% in 1998 to about 28% in 2001. Full-blown competition has acted as a strong catalyst for the development of mobile communications in China. Figure 5.3 shows the exponential growth of mobile subscribers in China since China Unicom entered the market in 1994. According to the latest statistics, both China Mobile and China Unicom are now among the top eight mobile operators in the world [14].

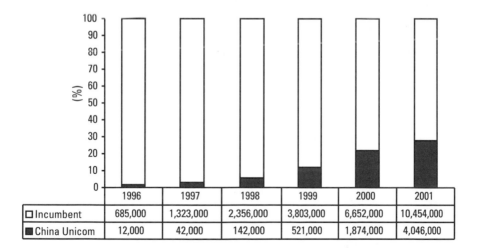

	1996	1997	1998	1999	2000	2001
□ Incumbent	685,000	1,323,000	2,356,000	3,803,000	6,652,000	10,454,000
■ China Unicom	12,000	42,000	142,000	521,000	1,874,000	4,046,000

Figure 5.2 Market-share growth of China Unicom.

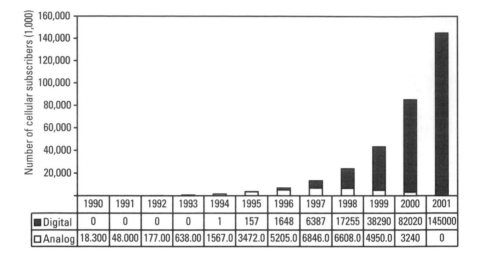

	1990	1991	1992	1993	1994	1995	1996	1997	1998	1999	2000	2001
■ Digital	0	0	0	0	1	157	1648	6387	17255	38290	82020	145000
□ Analog	18.300	48.000	177.00	638.00	1567.0	3472.0	5205.0	6846.0	6608.0	4950.0	3240	0

Figure 5.3   Growth of cellular subscribers in China. (*Source:* The MII.)

5.5   Emergence of a Fully Competitive Telecommunications Market

Another revolutionary step in the government's efforts in facilitating market liberalization was to split the former China Telecom in mid-1999 into four independent groups: China Telecom, China Mobile, China Satellite, and the Guo Xin Paging Company. Instead of following the approach of the United States when it vertically split up the Bell system into seven regional Bell operating companies according to geographical locations, the Chinese government adopted a strategy of horizontal separation—that is, China Telecom was split up into four groups according to specific services instead of geographic locations, and all groups were permitted to provide nationwide service. China Mobile is specifically dedicated to mobile phone services; China Satellite is similarly dedicated to the provision of satellite communications, while the Guo Xin Paging Company focuses only on radio paging services. Both China Mobile and China Satellite are financially and operationally independent, and the government clearly hopes that they will provide other services and compete with other operators in the future. The Guo Xin Paging Company was subsequently merged with China Unicom as a measure intended to enhance China Unicom's financial strengths, a process which is essential for listing China Unicom in the overseas stock market due to the huge customer base and long operational history of the former China Telecom's paging service. Additionally, two new operators—China Netcom and

China Railcom—have also been granted licenses since the establishment of the MII.

In August 1999, China Netcom was founded by four entities affiliated with the Chinese government: (1) the Chinese Academy of Sciences (CAS), (2) SARFT, (3) MOR, and (4) the Shanghai municipal government. The company closed its first round of private equity placements in February 2001, raising $325 million from a group of high-profile international investors, including News Group Corporation Digital Ventures, Goldman Sachs Private Equity, and other leading Hong Kong and Chinese financial institutions.

China Netcom's core competence lies in its strong shareholder support and a new generation of modern management. For example, many management team members of this company have overseas backgrounds and experience, which enables the company to operate in a totally different style compared with other state-owned operators. In 2000, China Netcom completed the first phase of the construction of a nationwide fiber optic backbone network, CNCNet, and a metropolitan access network, which pioneered the global deployment of advanced IP over DWDM transmission technology [15]. The company recently extended its last mile broadband access network to cover residential users using fiber local area network (LAN) technology.

China Railcom obtained a license in 2000 and officially went into operation by being spun off from the MOR on March 1, 2001. There were two prime reasons for its establishment. First, the railway industry was undergoing structural reform in which business conduct had to be detached from regulation, much like the telecommunications industry in the late 1990s. The second reason for establishing the new entity lay in the government's wish to extend competition in the telecom sector by utilizing MOR's private network. This network is the second largest in China after China Telecom. By the end of 2001, the total length of its toll circuits was 120,000 km covering more than 500 major cities in China. In the year since its establishment, it has accumulated a wealth of 13.6 billion yuan ($1.64 billion) in assets. Importantly, most employees working in the railway system were subscribers to this former private network, thereby constituting a significant customer base for China Railcom. The MII allows China Railcom to charge the customer 10% less than China Telecom. This asymmetric regulatory arrangement is undoubtedly favorable to the new entrant. At present, its core businesses are local fixed telephone service and Internet service. Its target is to obtain a market share of 5% in local fixed telephone service within 3 years.

The revised regulatory and industrial structures have dramatically changed the operational environment of the newly restructured China

Telecom, which was formally established in May 2000. Compared with its predecessor, the reconstituted company has been forced to relinquish its monopolistic status in the industry but, *pari passu*, has enjoyed much greater autonomy in strategic decision making and operational areas. At the same time, it faces increasing environmental uncertainties in both technological upgrading and market competition. For example, telecommunication networks are transitioning from circuit switching to packet switching while IP is widely applied in contemporary telecommunication systems. This implies that the networks that China Telecom has developed over the past years—constituting a burden of stranded investment—might become a handicap to its future development. Thus, China Telecom is currently at a crossroads in restructuring its future technology strategies. Although it has retained the entire network resources needed for local, long-distance, and international service, IP technology has enabled new entrants, such as China Unicom, China Netcom, and Jitong Telecom to compete effectively with China Telecom in providing long-distance and international services. By the end of 2001, in terms of IP telephony traffic in minutes, the market share of the four licensees was as follows: China Telecom (65%), China Unicom (20%), Jitong Telecom (10%), and China Netcom (5%) [16].

China Telecom is also facing competitive threats from mobile operators, such as China Mobile and China Unicom. Over the past 5 years, users of mobile services have been increasing at a higher growth rate than that of the fixed network. In 2000, the number of newly connected subscribers to mobile service overtook that of fixed networks. The clear threat that such developments pose lies in the potential of mobile to erode the customer base of the fixed phone network as both existing and new subscribers migrate and embrace mobile. Although China Telecom is currently providing Xiao Ling Tong service—a kind of mobile service with the Personal Handyphone System (PHS) technology—the relative weak technical strength of this technology (e.g. roaming function is not supported) makes it hard to attract high-end subscribers.

With the increase in new entrants and substitutes, China Telecom's bargaining power with suppliers and customers has been significantly decreasing. China Telecom was once the sole (monopsonistic) buyer of telecommunication equipment, enjoying a dominant bargaining position in concluding purchasing deals. There is now, however, more than one operator, so vendors can enjoy relatively higher bargaining power in concluding such sales contracts. China Telecom, therefore, can no longer rely on its free rider status in the field of equipment purchase linked to its most-favored monopoly status. Additionally, it faces a more uncertain and unpredictable cohort of consumers who are becoming more and more demanding and ever more conscious of their rights. They are now questioning the reasonableness of installation fees, surcharges, and high tariffs, and have increasingly begun to defend their interests through resort to law. As a result, China Telecom has come under unremitting pressure to improve customer service. The organization has been forced to adopt much faster responses to customers' demands and has had to be particularly fleet of foot in retaining the loyalty of business users who now have a greatly increased choice of several service providers, especially in long-distance and IDD services.

State-owned firms in transitional economies are frequently characterized as lacking financial and managerial resources [17]. This is certainly true of China Telecom. With decreasing backup from the government, the company has faced serious financial constraints in recent years. In addition, because of pressures from the public and the media, installation fees and surcharges have been significantly reduced. Between 1997 and 1999, the installation fee has been reduced from RMB 5,000 yuan ($625) to RMB 1,500 yuan ($180). On July 1, 2001, the MII formally announced cancellation of all installation fees and surcharges for telephone services. Another problem for China Telecom lies on the personnel front. It now exhibits a significant deficit in managerial and technique talent, as it has experienced a hemorrhaging of such talent to its newly arrived competitors.

To cope with environmental turbulence and resource limitations, China Telecom has reengineered its businesses to enhance its competitiveness across the wide array of telecommunication service. As all of its rivals have been newly established with modern corporate structures and well-focused business niches, their operational efficiencies have been relatively high. Against this burgeoning competitive background, examination of the organizational history of China Telecom reveals its strategy of dominating its supply chain together with a process of diversification. It has invested heavily in supplementary businesses (including designing, R&D, training, manufacturing, sales, and marketing) and affiliated businesses (including hostels, kindergartens, and cafeteria). When the new China Telecom was formally established in May 2000 after the split, it controlled in total more than 4,000 affiliated corporations. These affiliated corporations worked on nontelecommunications business, and the diversified operation proved to be inefficient and productive of dysfunctional consequences. In early 2001, China Telecom separated its core telecommunication business operations from its supplementary and affiliated businesses and established two independent corporations: the core business (telecommunications) corporation and the industrial corporation. Such rebrigading and restructuring of core activities has helped the company become "closer to the market" and has facilitated interorganizational cooperation and development.

For example, China Telecom's core business corporation outsources its market sales and system maintenance functions to the newly established industrial corporation. Thus, it can concentrate on network operation and compete with its rivals more effectively. At the same time, the industrial corporation can develop its businesses in the market rather than just providing supporting services to the core business corporation. For example, the training center of China Telecom in Guangzhou is located in Baiyun Mountain, a popular holiday resort. It has remained largely idle in the past, only occasionally providing training for internal employees. After the separation, this training center has become aggressively more businesslike and has begun to open its doors to the wider society, including China Telecom's competitive rivals, such as China Unicom and China Mobile.

Through this reengineering, the number of employees working on core businesses has been reduced from 530,000 to 397,000, and there are 140,000 employees working in China Telecom's industrial corporation [18]. To defend its dominance in the market, this core-business corporation has "stuck to the knitting" and kept up a trend of preemptive investment, mainly on broadband access and the backbone network. Simultaneously, it has extended its interorganizational liaison to various parties. For example, it has signed a corporate agreement with the Bank of China aimed at exploring the synergy between China Telecom's nationwide telecom network and the Bank of China's service network [19]. It has also signed a memorandum of understanding with NTT of Japan and Deutsche Telecom of Germany targeted on the achievement of greater cooperation [20]. In addition, China Telecom has established a Corporate User Department to deal more effectively with this critical market segment.

Compared with the situation from 1994 to 1998, the establishment of the MII implies a real institutional change, namely from monopoly or absolute domination of China Telecom to a level playing field conception of a competitive telecommunications market. For the first time in its history, China Telecom is now fully exposed on a competitive battleground and, in consequence, has begun to react more swiftly to market forces.

5.6 Conclusion

China's experience of telecommunications competition in the past few—critical—years clearly indicates that the establishment of a robust regulatory framework is a necessary and sufficient condition for the establishment and sustenance of fair competition and for the healthy future development both of new entrants and the incumbent. The past close affiliation between the incumbent operator and the regulator (the Chinese manifestation of regulatory capture in which the regulator is effectively made to conform to the agenda of the regulatee) forced new entrants into an extremely vulnerable position, while, to the contrary, the recent separation of regulatory and operational functions in the restructured MII has strengthened the independence of its regulatory arm and led to the revival of a new entrant, such as China Unicom.

With the policy transfer of regulatory models from abroad and their continuing incorporation into the Chinese telecommunications policy milieu, the incumbent has begun to feel the pressure of competition and, in consequence, react more swiftly to market forces. While the vestiges of bad old ways and habits still continue to inhabit China Telecom—revealed, for example, in the presence of an inflated number of noncore affiliates—evidence of new ways of thought in the incumbent are present in increased labor shedding as the organization attempts the difficult process of "rightsizing." The conclusion seems inescapable that the incumbent has been forced to seriously change its structure and working practices as a consequence of greater contestability. A prime role of the regulator is to act as an accelerator of such contestability.

Although the MII has still not attained the status of an independent regulator in the full sense of the term, its quasi-independent status has brought dramatic changes to the telecommunications industry. Such a revolution (admittedly unfinished) in the construction of the building blocks of a reregulated telecommunications policy framework will surely presage further liberalization of the Chinese telecommunications policy sector in the future. The lesson of recent Chinese telecommunications history is instructive: an ineffective and essentially immature regulatory framework has proven costly for China. However, having acknowledged that the regulatory issue continues to exist as unfinished business on the Chinese telecommunications policy agenda, it should readily be acknowledged that moves already taken to strengthen the regulator, most notably through MII reorganization, have ensured that regulatory authority can be effectively exercised over telecommunications players to the evident benefit of the wider community. The Chinese regulator, like its counterparts in so-called first mover countries should not be a mere "watchdog with rubber teeth." Recent developments suggest that realization of this truth is well implanted in Chinese telecommunications public policy fora. Much has already been achieved to provide a corrective to the statist and monopolistic practices of the past, which have undoubtedly proved costly to China's modernizing ambitions. Much undoubtedly remains to be achieved in further regularizing the position of the regulator and ensuring that the office safeguards and ensures the effective continuation of the bold Chinese telecommunications policy experiment.

References

[1] Selvarajah, K., "Key Challenges for Development of Telecommunications in the Asia-Pacific Region," *APT J.*, May 2000, pp. 13–17.

[2] WTO, "Reference Paper on Regulatory Principles," http://www.wto.org, 1997.

[3] Selvarajah, K., "Telecommunications Transition in Sri Lanka: A Model for Small Countries?" *Info*, Vol. 1, No.1, 1999, pp. 77–93.

[4] ITU, America's Telecommunication Indicators, 2000a.

[5] Cai, Y., "The Impacts of Information Infrastructure on Telecommunications Regulation," *Proc. Int. Conf. Information Infrastructure,* Beijing, China, April 1996, pp. 25–28.

[6] Tan, Z. A., "Regulating China's Internet: Convergence Towards a Coherent Regulatory Regime," *Telecommunications Policy*, Vol. 23, 1999, pp. 261–276.

[7] Wang, X. D., *Informationization: Choice of China in the 21st Century*, Beijing, China: China Social Science Publication House, 1998.

[8] Yun, T., "By Whom, the Ministry of Posts and Telecommunications or the Ministry of Radio, Film and Television?" *China Computer World*, December 15, 1997.

[9] Luo, G., Report to Ninth National People's Congress on Institutional Restructuring, March 1998.

[10] Wang, X. D., and Y. Xu, "The Reform of the Regulatory Framework and Its Impact on the National Information Industry," *Communications Weekly*, July 8, 1998.

[11] "MII Published Authorized Size," *China Communications News*, Vol. 2, No. 6, 1998.

[12] Xu, Y., and D. C. Pitt, "Competition in the Chinese Cellular Market: Promise and Problematic," in *The Future of the Telecommunications Industry—Forecasting and Demand Analysis*, D. G. Loomis and L. D. Taylor (eds.), Boston, MA: Kluwer Academic Publishers, 1999, pp. 247–264.

[13] Lovelock, P., "Case Study of IP Telephony in China," http://www.itu.int/iptel, 2000.

[14] Xu, Y., "Return of the Tigers: Asian-Pacific Innovation in Mobile Communications," *Info*, Vol. 3, No. 3, 2001, pp. 231–242.

[15] China Netcom, "Introduction to China Netcom," http://www.cnc.net.cn, 2001.

[16] China Telecom, "The Growth Comes from Competition," *Telecommunications Market Information Weekly*, No. 146, 2002, pp. 3–5.

[17] Peng, M. W., and S. P. Heath, "The Growth of the Firm in Planned Economies in Transition: Institutions, Organizations, and Strategic Choice," *Academy of Management J.*, No. 21, 1996, pp. 492–528.

[18] Chang, X. B., "China Telecom Stands at a New Starting Point," *People's Daily*, March 24, 2001.

[19] "China Telecom Shares Network Resource with Bank of China," *China Communication Net*, May 24, 2001.

[20] China Telecom, "China Telecom's Ambition Is Not Limited to Being No. 1 in the Domestic Market" http://www.chinatelecom.com.cn, January 2, 2001.

6

3G Licensing in China: Telecommunications Policy Formulation in a Broad Economic Context

Like the former Ministry of Posts and Telecommunications, the MII was designated by the Chinese government as a regulator not just for telecommunications operations, but also for telecommunication equipment manufacturing. This policy decision implies that the Chinese regulator has had to consider the broader economic context when formulating telecommunications policy. This regulatory stance, effectively ensuring that the regulator is a player in the field of Chinese industrial policy, has rendered it difficult for the MII and the former MPT to take a technologically neutral regulatory position, such as that practiced by some other overseas regulators like the FCC of the United States. In fact, using the domestic telecommunications market to support the development of domestic equipment manufacturers (as revealed later in this chapter) has been a consistent policy approach adopted by the Chinese government. This is particularly evident in the case of the burgeoning Chinese mobile telecommunications marketplace.

6.1   Chinese Telecommunications Manufacturing Industry

When China began to reform its telecommunications system in the early 1980s, its policy-making elite was quick to realize that its infrastructure lagged far behind that of developed economies, not just in terms of teledensity but also in terms of technological sophistication. The entire network was based on analog technology, and Chinese vendors could only produce switching systems based on increasingly obsolete cross-bar and step-by-step technologies. Both long-distance and international calls had to be connected manually via human operators.

In 1982, Fujian province imported and installed the first State Planning Commission switching system in China. The high quality and innovative features of the Stored Program Control (SPC) system consequently brought about a boom in equipment imports. Because of its late mover advantage, Chinese operators immediately caught up with the latest technologies, including SPC switching, digital microwave, optical fiber, and satellite communications. By the end of 1998, 99.6% of the transmission system and 98.8% of the local switching system had been digitized. At the same time, however, billions of U.S. dollars were spent on imports.

In order to enable domestic vendors to upgrade their technology, thus reducing dependence on foreign products, the government in the early 1980s formulated a four-step strategic policy: import, digestion, absorption, and creation. Using the domestic market to attract foreign technology has been a preferred strategy for upgrading the technology of domestic manufacturers. The huge telecommunication market has provided the Chinese government with strong bargaining power with which to urge foreign vendors to transfer their technology during bilateral trade deal negotiations, par- ticularly during the process of establishing joint venture arrangements. For example, when the first joint venture for switching equipment— Shanghai Bell, a joint venture between Alcatel's Belgian branch and the MPT— was established in 1984, the Chinese government defined precise terms on which technologies should be transferred to the Chinese side by the Belgian partner and in what manner [1]. In 1998, more than 74% of Shanghai Bell's hardware and 90% of its software products were developed and fabricated in China. The government allocated quotas to this joint venture, which defined the maximum that Shanghai Bell could sell to the domestic market. Its products have subsequently been exported, inter alia, to Korea, Vietnam, the Philippines, Germany, Spain, Australia, and Belgium. Between 1984 and 1999, its total revenue had reached 28.20 billion yuan ($3.42 billion) [2].

In addition to establishing joint ventures, the Chinese government has provided favorable support to domestic manufacturers. This support has included the assignment of a research grant for R&D, low interest loans, discounted tax rates, and a generous provision of land in high-tech industrial parks. Such support has been extremely effective. By the mid 1990s, several domestic manufacturers were well established, including Julong, Datang, Zhongxing (ZTE), and Huawei. In China, professionals in the telecommunication industry use the title Ju Da Zhong Hua (the combination of the first character of the above corporations' names), which in Chinese stands for Giant China, to represent the domestic telecommunication manufacturing industry.

This constellation of domestic vendors has achieved significant success in the manufacturing of transmission and switching equipment. In 1992, the first SPC switching system of 10,000 lines was manufactured in China with China-owned intellectual property rights. Since then, under government coordination, domestic manufacturers began to dominate the local market. For instance, in 1998, domestic manufacturers supplied 98% of the newly installed switching equipment for fixed local networks and 50% of newly installed optical transmission system. In 1999, the market share of local vendors in newly installed switching equipment rose to more than 99%. At the same time, more and more local products have been exported to foreign countries. In 1999, the total revenues from exporting telecommunication equipment and systems reached $4.66 billion [3].

This achievement by domestic manufacturers has strongly encouraged the Chinese government and industrial sectors to move aggressively into the prosperous mobile communication sector. In January 1999, the State Council issued a document entitled "Several Issues on Speeding up the Development of the Chinese Mobile Communications Industry." According to this document, from 1999 to 2003, the government intends to use 5% of the telephone installation fees as a special grant for R&D on mobile technologies. In the 1999–2000 fiscal year, the Ministry of Finance raised 580 million yuan ($69 million) to support 31 projects in seven categories relating to key mobile communication technology. By the end of 1999, total investment on R&D in mobile technology sourced from both the government and industry reached 4.5 billion yuan ($545 million) [4].

In November 1999, the first mobile handset assembly line began operations. In the same year, domestic vendors made a breakthrough by entering the mobile equipment market. In 2000, 10 domestic vendors produced 3.44 million handsets, of which they sold 3.26 million. The total capacity of the mobile switching system sold in 2000 reached 5.85 million lines while sales of mobile base stations reached 145,000 sets. Figure 6.1 shows the growth in market share of Chinese domestic vendors from 1999 to 2001.

Domestically manufactured mobile products have intensified competition in the equipment market. In response, foreign vendors were forced to reduce their prices. According to the MII, as a result of this competition in price, Chinese mobile operators have saved a total of 15 billion yuan ($1.82 billion) over the past 2 years [4].

Until recently, Chinese manufacturers successfully obtained patents for SIM cards, batteries, LED monitors, filters, high-frequency switches, and other components. Foreign GSM vendors, however, have patented most of the key GSM technologies. According to recent statistics, there are 140 GSM patent categories and 369 independent patents, all of which are owned by only 17 corporations. The patent fee ranges from 3% to 30% of revenue [5]. Thus, Chinese domestic vendors have to pay onerous patent fees to foreign companies.

Unlike GSM, Chinese vendors moved very fast in code division multiple access (CDMA) technology. So far, Chinese domestic manufacturers have developed and patented many key switching and base station technologies for the CDMA system with the exception of the baseband chip, which has been patented by Qualcomm of the United States. This includes CDMA IS-95A

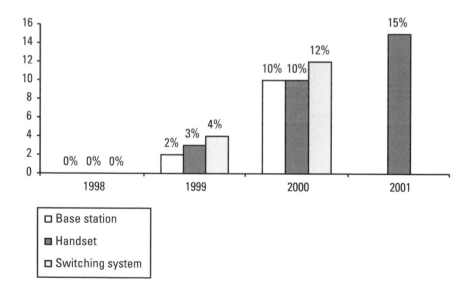

Figure 6.1 Growth of market share of Chinese domestic vendors in mobile product supply. (*Source:* MII.)

and CDMA2000-1X switches, base stations, and handsets. Huawei, for example, has invested heavily in the development of CDMA technology since 1996 and has obtained 97 patents. Currently, it is able to fulfill a regular order for CDMA systems in 45 days and an emergency order in 30 days [6].

The success of Chinese vendors in terms of CDMA technology has strongly encouraged the Chinese government to roll out a CDMA network. On January 1, 2001, Great Wall, a joint venture between China Telecom and the PLA, which has itself provided a CDMA service in Beijing, Shanghai, Xian, and Guangzhou without owning a formal license, was transferred to China Unicom. This transference sent a clear signal that the Chinese government would like to boost the CDMA network, despite the fact that Great Wall's network had only 200,000 subscribers, substantially less than the 56 million subscribers to the GSM system in the late 2000 [7].

On January 13, 2001, China Unicom announced that it would construct the first phase of the world's largest CDMA network with a capacity of 13.3 million lines. By 2003, the total capacity of the CDMA network is predicted to reach 30 million. On the next day, Unicom Horizon, an independent subsidiary of China Unicom, obtained a license for operating a CDMA network. China Unicom also signed the "CDMA Intellectual Property Rights Framework Agreement" with the U.S. company Qualcomm. Under this agreement, equipment suppliers of Unicom Horizon were only obliged to pay Qualcomm intellectual property rights (IPR) royalties at a rate of 1.5%. This is a very favorable arrangement, given that Qualcomm charges Korean manufacturers IPR loyalties of 6% [interview with ZTE Corporation, April 27, 2001].

An important political factor also underpins the Chinese government's decision to deploy CDMA technology. In fact, one of the key bargaining chips of the Chinese government in negotiation with the United States on China's accession to the WTO was the issue of whether or not China should use Qualcomm's CDMA technology. In the event of GSM technology dominating most cellular communications markets in the world, the U.S. government was very keen to use this opportunity to pave the way for Qualcomm's CDMA technology to penetrate into the world's largest mobile phone market. This was the ostensible reason why the Chinese government had never given a concrete signal that CDMA technology would be deployed until China reached agreement with the United States on its accession to the WTO in 2000. Bearing in mind that Chinese vendors also obtained certain intellectual property rights on CDMA, this was a win-win deal.

A network with nationwide potential connoted a new giant market brimming with commercial possibilities. Between March and April 2001,

China Unicom invited tenders for CDMA network switching systems and base stations worth 20 billion yuan ($2.42 billion). This resulted in 12 companies bidding for switching systems and six for base stations. Realizing that the Chinese government would like to use CDMA to stimulate the domestic manufacturing sector, almost all foreign manufacturers set up partnerships with domestic vendors or China-based joint ventures: Huawei (Motorola, Ericsson), Shanghai Bell (Samsung), Nanjing Panda (Ericsson), Qingdao Lucent (Lucent), Guangdong Nortel (Nortel), Julong (Ericsson), Shouxin (LG), Datang (Nortel, Lucent), Hangzhou Motorola (Motorola), and Eastern Communication (Motorola, Ericsson). Only Zhongxing (ZTE) joined the bidding independently without setting up a partnership with foreign vendors. Clear evidence of support for domestic manufacturers was the example of ZTE—set to bid for 6.6 million lines out of 13.3 million lines by Unicom Horizon, the highest among all bidders.

In May 2001, China Unicom formally released the results of the tender. ZTE and nine other companies won the bid and formally signed an agreement with Unicom Horizon. ZTE won 7.5% of the market, agreeing to supply a CDMA system of 1.1 million lines to Guangdong, Guizhou, Yunnan, Shanxi, Hainan, Hubei, Inner Mongolia, Ningxia, Qinghai, and Tibet. Although ZTE complained that its market was mainly located in remote areas [interview with ZTE Corporation, April 27, 2001], it was nevertheless a milestone for the Chinese telecommunication manufacturing industry. For the first time a Chinese domestic vendor was able to supply a total network solution to mobile operators.

Chinese manufacturers, however, are unlikely to rest on their laurels. Time division synchronous code division multiple access (TD-SCDMA), the 3G standard proposed by China and approved by the International Telecommunication Union (ITU), has extended the ambitions of the Chinese manufacturing industry beyond second-generation (2G) mobile communications. Flourishing developments in short message and other mobile data services in the contemporary Chinese mobile market have further boosted their ambitions.

6.2   The Dynamics of Mobile Data Communications in China

With the help of market liberalization, mobile communications in China have seemingly enjoyed impressive growth in terms of the number of users. An increase in users, however, does not imply correlation with use. Figure 6.2 shows the decline of average revenue per user (ARPU) per month and minutes of usage (MOU) per user per month of China Mobile (HK)

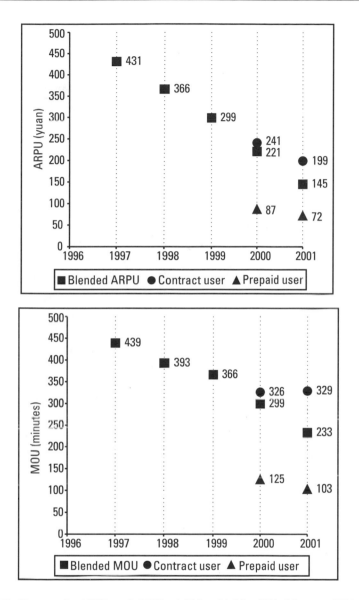

Figure 6.2 Changes in ARPU and MOU of China Mobile (HK). [*Source:* 2002 Annual Report of China Mobile (HK).]

over the last 5 years. The ARPU was 221 yuan ($26.69) in 2000, representing a decline of 26.1% compared with 1999. The MOU in 2000 was 299 minutes, a decline of 18.1% compared with 1999. In 2001, the ARPU and MOU have been further reduced to 145 yuan ($17.51) and 233 minutes.

According to China Mobile, the decline in ARPU and MOU was mainly due to the substantial growth in lower usage subscribers and, in particular, subscribers of prepaid services.

In order to fully exploit network potential and generate more revenues from the current subscriber base, both China Mobile and China Unicom have introduced a bundle of value added services in recent years. Services offered include caller number display, voice mail, short messages, call forwarding, call waiting, three party calling, and Voice over IP (VoIP) long-distance calls. In order to keep pace with mobile commerce developments elsewhere in the world, on World Telecommunication Day 2000 (May 17th), both China Mobile and China Unicom formally launched their nationwide wireless application protocol (WAP) services. Currently available WAP services include mobile banking, stock trading, news, weather reports, and e-mail.

Such services, however, have not yet been taken up by subscribers. In Beijing, 5 months after the service was first launched, China Mobile only managed to sign up 8,000 WAP users [8]. A survey during the period from December 2000 to February 2001 revealed that only 2% of Chinese subscribers access mobile Internet via WAP phones, the lowest among all eight Asia-Pacific economies studied (Figure 6.3).

In contrast, subscribers to the NTT DoCoMo's i-mode service have increased steadily in Japan at an average rate of around 668,700 per month

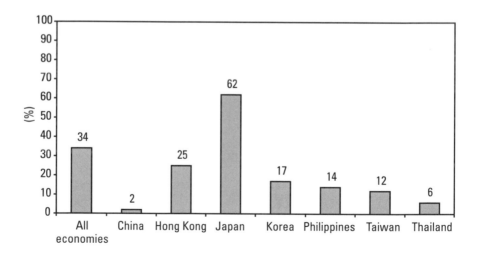

Figure 6.3 Mobile Internet access using WAP phones. (*Source:* http://www.tnsof-res.com/apmcommerce.)

since the service was launched on February 22, 1999. In May 2000, the operator had to temporarily suspend the sign-up of new subscribers because demand surpassed system capacity. By November 2001, the total number of i-mode subscribers had reached 39.2 million [9].

One of the major factors behind the contrasting response to WAP and i-mode from the market could result from the fact that NTT DoCoMo has created a successful business model. It provides a platform upon which providers supply content, either for free or for a small premium fee, which is set at a maximum of 300 yen per month. NTT DoCoMo shares traffic revenue related to content with the providers of the content, and typically keeps a 9% commission fee. This has provided a strong incentive for content providers. As a result, a large amount of Japanese content has been available. In April 2000, for example, 448 application alliance partner companies and 8,023 voluntary i-mode Internet Web sites, including 20 search engines, had content provision contracts with NTT DoCoMo. The availability of such impressive content has attracted more and more subscribers. As more subscribers sign up for i-mode, more content providers are enticed to provide more content. A positive feedback process is thus established [10].

NTT DoCoMo's success in Japan has encouraged Chinese cellular operators to set up a similar business model. In November 2000, China Mobile introduced the Monternet program. Under this program, service providers can access the carrier's mobile network at any place to provide nationwide service. This is also known as the "one-stop shop, China-wide service" arrangement. China Mobile keeps 9% of the traffic revenues, and the information service providers receive 91% of the revenue. If the arrangement is to include coverage for bad debts, China Mobile increases its commission to 15%.

The Monternet program has generated an overwhelming response from service providers. By the end of March 2001, 102 service providers had joined the Monternet program for cooperation in the mobile Internet market. As there are so many content providers, China Mobile has to cherry-pick in order to supply its limited capacities to the most valuable content providers. These service providers include Sohu, Sina, and other popular Chinese Internet portals. None of these companies have, to date, made any profit through their Internet businesses, but Monternet has opened up new possibilities, with subscribers paying for every message they receive.

Currently, these service providers offer several types of services, including message-on-demand, message broadcasting, and stock trading. For example, subscribers can visit the Web site of Sohu and subscribe to customized news, such as sports and entertainment. This allows them to receive the latest news via their handsets on a regular basis, say three times a day. At present, China Mobile charges 0.2 yuan ($0.02) for each piece of news. For information ordered by the handset, China Mobile will charge a different rate for terminating different kinds of messages (see Table 6.1 for an overview of charges). In addition to the transmission fee, content providers may also charge a content fee, and the rate varies depending on the individual service provider. For instance, NewPalm delivers daily weather reports to a subscriber's handset for a total monthly fee of 4 yuan ($0.48). China Mobile collects this content fee on behalf of the content provider, and also shares the transmission fee with them.

In order to facilitate the Monternet program, China Mobile set up a subsidiary under the name of Aspire in the last quarter of 2000. Hewlett Packard invested $35 million in the company and owns 7% of it. On January 9, 2002, Vodafone made an investment of $34,965 million for a 9.99% equity stake in Aspire. Aspire is currently involved in the construction of the Mobile Information Service Center (MISC) platform. The MISC is meant to serve as the common platform for China Mobile's entire mobile Internet services. A unified MISC platform will provide mobile subscribers with mobile data roaming capabilities. The MISC will also provide a uniform data interface open to third-party service providers, through which standard network information (such as billing) can be provided. The segregation of service platforms from the basic mobile communication services will ensure that all mobile communications networks developed through the platform can be smoothly migrated when they are upgraded to 2.5G and 3G, making them truly "future compatible networks" [11].

Table 6.1
Termination Rate for Messages of Different Content (per Piece, Yuan)

Stock	0.20
Weather	0.10
Flight	1.00
Foreign exchange	0.10
Train	0.30
News	0.20
Dictionary	0.10

Source: China Mobile.

Another strategy for facilitating China Mobile's Monternet program is the upgrade of its current circuit-switching network to a packet-based one. On January 21, 2001, China Mobile formally launched its General Packet Radio Service (GPRS) network project. Since May 17, 2002, GPRS service has been available in 160 Chinese cities.

Despite concerted efforts on the part of operators to upgrade their technology, the real growth of mobile data services actually lies in the short message service (SMS)—one of the simplest value-added services based on the GSM standard. For China Mobile (HK), the usage volume of SMS increased from 126.7 million messages in the first half of 2000 to 4.776 billion messages in the second half of 2001, representing an average compound half-yearly growth rate (CHGR) of 235% (Figure 6.4). On Chinese New Year's Day on February 12, 2002, more than 100 million short messages were delivered over China Mobile's network, which brought in revenue of around 10 million yuan ($1.21 million) within a single day.

This phenomenal increase in SMS use is hardly surprising. According to market research sponsored by China Mobile, the mobile data service most in demand by Chinese users is e-mail (Figure 6.5) [12]. This reflects the special characteristics of China's infrastructural configuration. By the end of 2001, there were 145 million mobile subscribers but only 30 million Internet subscribers. This implies that the wireless Internet is more widely accessible than wireline Internet. Thus, subscribers use their mobile handsets to substitute for the PC for sending and receiving e-mail, or, more accurately, text

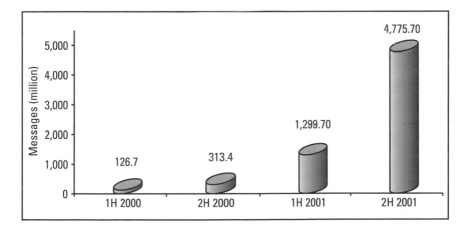

Figure 6.4 SMS usage volume of China Mobile (HK) in 2000–2001. [*Source:* 2002 China Mobile (HK) Annual Report.]

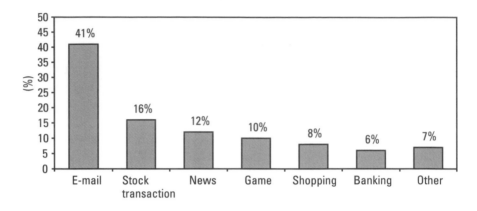

Figure 6.5 Demand for mobile data services in China. (*Source:* [12].)

messages. In fact, some wireline Internet portals, such as http://www. etouch.com, provide a service that enables users to send SMS via their PC to the handsets of mobile phone users. In this way, the simple SMS acts as a bridge between the wireless Internet and the wireline Internet.

Fieldwork conducted in May 2001 by one of the authors revealed that SMS has the following advantages in the specific context of the Chinese telecommunication market:

- SMS provides an optimum economic solution for communication. The end user price for sending and receiving a message is 0.1 yuan ($0.012), whereas a 1-minute call costs 0.40 yuan ($0.048). In addition, the minimum charged unit for mobile telephony in China is 1 minute. This implies that for unsophisticated information, the short message is more cost-effective. And if the two communicating parties are located in two different cities, the advantage is even more evident, since the long-distance charge for mobile service is 0.70 yuan ($0.084) per minute.

- SMS can be used in certain special circumstances. For example, when one of the two parties is in the middle of a conference, he can still receive and reply to short messages without disturbing other conference participants. Wide application of the short message system might significantly modify the phenomenon, familiar in Asian contexts, of talking on the phone in conference rooms and cinemas.

- The short message can express some information more adequately than the verbal medium. A large number of messages, for instance, have been delivered as greetings during Chinese New Year, as noted, and Valentine's Day.

- SMS is more suitable for broadcasting information. Many m- commerce companies, like Realvision, began to provide solutions for some companies to broadcast internal corporate information. Organizations like insurance companies, hydroelectric companies, and national and city police forces, which have an array of dispersed branches and employees, are heavy users of the short message broadcasting service.

- Because of important inhibiting cultural factors, Chinese users are reluctant to leave voice messages but prefer to leave text messages, the perfect medium for which is SMS.

- According to some China Mobile managers, the short message has become a type of new and informal literature. Featuring political jokes and adult humor, for example, this medium effectively bypasses the strict control that continues to be exercised by the government over the public media.

The enormous growth of SMS services reveals the huge potential of the wireless data communication market in China. As a result, many players have moved into this market. At the same time, new technical solutions and business models have been created. In fact, the only bottleneck, according to Ken Woo, vice president of Cyber-on-Air (an active player in the Chinese mobile commerce market), lies in the slow transmission speed of the network. Developing a broadband 3G network is therefore critical to the future development of mobile data communications in China. Experience in foreign countries, however, has shown that, because of the broadband characteristics of 3G technology, spectrum allocation is normally a major concern of the regulator in issuing 3G licenses.

6.3 Spectrum Regulation in China

The radio spectrum is an enormously valuable and scarce natural resource. To fully exploit the potential of this resource efficiently, an increasing number of countries are adopting the stance of commercializing the radio spectrum. In 2000, the licensing of 3G mobile generated a spectrum auction fever in Europe. The $47.5 billion license fee in Germany and $33 billion license fee in the United Kingdom encouraged governments in other countries to follow the same approach in the expectation of obtaining similar windfalls. Negative stock market reaction, however, has led to a decline of auction fees in later-mover countries (Figure 6.6).

In China, according to the policy statement "Radio Spectrum Regulation of the People's Republic of China" published in 1993, one of the four principles of radio spectrum regulation is that the users should pay for occupying spectrum. At present, for cellular mobile services, both China Mobile and China Unicom have obtained spectrum at no charge, but individual subscribers are required to pay a so-called spectrum occupation fee every year. The spectrum occupation fee must be handed over by the operators to the Radio Regulatory Department of the MII, which was formerly known as the State Radio Regulatory Commission. At present, the spectrum occupation fee is 50 yuan ($6.06) per subscriber per annum. Taking 2001 as an example, spectrum occupation fees reached a total of 7.25 billion yuan ($878.79 million), against a total of 145 million mobile subscribers by year's end.

So far, calls for a spectrum auction are not as strong in China as in other countries even though the newly published Telecommunications Regulations of the People's Republic of China in 2000 included auctions as one of the appropriate methods of allocating radio spectrum and other telecom resources. The relevant articles of the Telecommunications Regulation of 2000 are as follows:

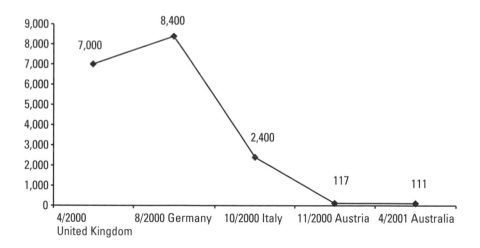

Figure 6.6 Global trend of 3G auction price. (*Source:* Briefing paper of ITU 3G Licensing Workshop, http://www.itu.int/3g.)

Article 27: The State applies unified planning, centralized administration, and rational allocation of telecommunications resources, and institutes a system of paid use.

"Telecommunications resources," to which reference has just been made above, refer to limited resources used for realizing telecom functions, such as radio spectrum, satellite orbital positions and telecom network numbers.

Article 28: Telecommunications operators should pay telecommunications resource charges for their occupation and use of telecommunications resources. Charging procedures will be formulated by the authority responsible for the information industry in conjunction with the authorities in charge of finance and prices under the aegis of the State Council, and shall be promulgated for implementation after approval conferred by the State Council.

Article 29: Telecommunications resources will be allocated with full consideration given to issues of planning, usage and the forecasted capacity of the related resource.

The telecom resource can be allocated in the way of assignment, or, in the way of auction.

This regulatory green light, however, has not led to an immediate and sudden boom in telecommunication resource auctions. Thus far, auctions have only been used in some provinces for allocating telephone numbers containing culturally lucky numerals, such as 8. Spectrum auctions have, to date, not occupied a position on the regulator's agenda. On September 11, 2000, Wu Ji-chuan, the Minister of Information Industry, told a journalist during an interview with the *Financial Times* that the 3G auction in Europe is inapplicable to the Chinese market, saying that he "[did] not wish to see developments in 3G technology generating a 'bubble effect'." Clearly exercised at the possibility that supply could far exceed demand, he pointed to the German experience and posed the question, "how many years will it take the licensees to retrieve the $4 billion license fee in a context in which no definitive standards and array of feasible services for 3G services have been fully defined?" [13].

In early March 2001, Minister Wu further highlighted the problematics of 3G licensing when he was interviewed by a group of overseas journalists, saying, "you should analyze why investors keep selling telecom shares in Europe and the Americas. The 3G licenses are too expensive. This will undoubtedly jeopardize operators' positions." He emphasized that 3G licenses in China would be issued via a smoother and more efficient process, taking into account the specific context of China. Speculative purchase of licenses would be discouraged [13].

In addition to concerns about the financial burden on operators, continuing state majority ownership of China Mobile and China Unicom constitutes another disincentive for auctioning the spectrum. Within the corset of state ownership, auctions effectively connote a situation in which state assets are merely rebrigaded from one pocket of the State Treasury to another. Unlike countries with a strong private sector tradition, the benefits to government of auctions are not clearly in evidence. They are effectively much less attractive as allocative mechanisms than in economies in which telecommunications operations are located totally or partially in the hands of the private sector (interview with Department of Policy and Regulations, MII, April 25, 2001).

During the above-mentioned interview, however, Minister Wu confirmed—significantly—that operators should pay for the spectrum. A payment method would be designed in such a way as to meet the specifics of the Chinese situation. This departure is consistent with the principle enunciated in the Telecommunications Regulations of 2000 specifying that "telecommunications operators should pay telecommunications resource charges for their occupation and use of telecommunications resources" [14]. According to the MII, the purpose of this principle was to prevent operators from hoarding such resources without effectively using them, or at best making limited use of them. Either would be productive of allocative inefficiencies. By allocating spectrum according to commercial accounting practices, supply and demand could be appropriately adjusted and aligned, while resources could thus be more effectively allocated [15].

This policy innovation marks a significant departure from current practice. As mentioned earlier, it is individual subscribers who now pay the spectrum occupation fee rather than the operator. In consequence, operators have not been subject to any pressure to improve spectrum efficiency. For example, if China Mobile only had one subscriber, the government would receive only 50 yuan for radio spectrum occupied by China Mobile, even though this spectrum can accommodate 100 million subscribers. Another disadvantage of the past arrangement lay in the fact that operators were not directly obliged to pay for spectrum and, additionally, were typically reluctant to assist the regulator in collecting the spectrum occupation fee. Accordingly, this public resource was not effectively compensated. In this sense, 3G licensing formally marks the beginning of a process of spectrum commercialization in real terms. The resultant shift in payment obligation from the subscriber to the operator should presage greater operational allocative efficiency.

According to a senior director of the Radio Regulatory Department of the MII, the proposed new method will mean that the regulator defines the price of the spectrum by benchmarking a reasonable economic price and then allocating the spectrum to operators according to this defined rate. Although this process is vulnerable to the influence of subjective factors, in theory it could prevent the phenomenon of overspeculation of the spectrum [interview with Radio Regulatory Department, April 26, 2001].

In June 2001 the MII promulgated a tender invitation for spectrum in the 3.5-GHz range, which is specifically used for wireless local loop services. The prices to be offered for spectrum were not included in the criteria, but the basic capability of the bidder and his spectrum usage schemes were to be major features of this "beauty contest" [16]. According to an official from the Radio Regulatory Department, the driving force behind this approach is the presence of a large number of interested parties (including the army) who would like to obtain this spectrum. To enhance the fairness and transparency of the licensing procedure, and to avoid any potential controversies, the regulator firmly supported this objective mechanism [interview with Radio Regulatory Department, July 26, 2001]. The tender process marked a major breakthrough in spectrum regulation in China, as it indicates a major renunciation of the older practice of bureaucratic assignment of scarce resources.

On April 28, 2002, the State Planning Commision, the Ministry of Finance, and the MII issued a joint statement announcing the cancellation spectrum occupation fee from July 1, 2002. In the mean time, a standard spectrum usage fee for GSM networks was published. The standard spectrum usage fee for GSM networks would be adjusted progressively over a period of 3 years starting from July 1, and the adjustments would be effective for a total of 5 years.

The operators have been burdened with high financial pressure and are liable to pay the spectrum usage fee directly to the regulator. China Mobile claimed that the fee arises would cost it an estimated 224 million yuan, and 462 million yuan after taxes in the first, second, and third years, respectively, after the adjustments take effect on July 1, 2002. China Unicom said in a statement that it estimates the fee hikes will increase its expenses on an after-tax basis by 64.32 million yuan in the first year, 211.72 million yuan in the second year, 292.12 million yuan in the third year, 318.92 million yuan in the fourh year, and 345.72 million yuan in the fifth year.

Bear in mind that 3G is emerging in the next one or two years. This 5-year spectrum usage fee scheme indicates that the auction will not be applied in China for allocating 3G spectrums.

In fact, unlike Hong Kong and other economies, the principal issue for 3G licensing appears to lie less in the issue of spectrum allocation and more in the appropriate choice of 3G standard for Chinese operators. 3G is doubly significant—not simply for Chinese operators but also for the domestic manufacturing industry.

6.4   TD-SCDMA: The Goose That Lays Golden Eggs?

On June 29, 1998, the last day set by the ITU for the submission of 3G standards from individual member countries, China faxed its own proposal to the headquarters of the ITU in Geneva. Signed by the minister and two vice ministers of the MII, this proposal outlined a so-called time division synchronous code division multiple access (TD-SCDMA) standard. By the set deadline, the ITU had received a total of 16 proposals from North America, Europe, Japan, and China.

After much discussion and debate, TD-SCDMA, together with SC-TDMA (UMC-136), MC-TDMA (EP-DECT), MC-CDMA (CDMA 2000), DS-CDMA (WCDMA), were accepted as radio interface standards by the ITU-R on November 5, 1999. In May 2000 at the World Radio Conference (WRC) of the ITU, TD-SCDMA was accepted as one of the three 3G transmission standards, together with CDMA2000 of the United States and wideband CDMA (W-CDMA) of Europe. On March 16, 2001, TD-SCDMA made another breakthrough: at the 11th plenary session of the Third Generation Partnership Project (3GPP), the entire technical scheme of the TD-SCDMA standard was accepted by 3GPP and was subsequently included in 3GPP's 4th Release. This breakthrough marked the acceptance of TD-SCDMA not only by the ITU, but also by a powerful industry alliance of operators and vendors [17].

TD-SCDMA marks a hugely important milestone for the Chinese telecommunications industry, as it is the very first telecommunication standard proposed by China to gain international acceptance. The long-term proactive efforts of the Chinese government and the telecommunications industry to gain credibility on the global telecommunications stage had finally paid off.

As early as June 1992, the SPC and the former MPT sponsored a research project entitled Digital Mobile Communications Technology (GSM). The Chinese Academy of Telecommunications, which was later renamed the Datang Telecom Technology & Industry Group in 1999, undertook this project. After 4 years of intensive work, the prototype of the GSM system passed the test of the SPC and the former MPT. However, because of early-mover advantage, foreign vendors remained in a position of dominance within the Chinese GSM market. As Figure 6.1 indicates, Chinese vendors have still only succeeded in securing a limited market share. At the same time, they have to pay enormously high patent fees to foreign intellectual property rights owners. According to recent statistics, such payments had reached over 10 billion yuan ($1.21 billion) by the middle of 2000 [18].

In 1997 the ITU called for proposals for IMT-2000 (3G) standards. This provided the Chinese telecommunication manufacturing industry with a significant opportunity. In the same year, Datang's SCDMA system—a wireless access technology—passed the state-level assessment. As it adopted many leading-edge technologies (such as synchronous CDMA, software-defined radio, and smart antennae) the former MPT decided to propose the Chinese 3G standard on the basis of SCDMA. In May 1998 a group of engineers began to draft the proposal for TD-SCDMA, which was later submitted to the ITU [19].

Compared with WCDMA and CDMA2000, TD-SCDMA has certain distinguishing characteristics [20]. First, the application of smart antennae and the low chip rate of 1.28 Mbps can significantly improve the efficiency of spectrum usage. For example, in order to transmit traffic at the rate of 2 Mbps, both CDMA2000 and WCDMA need two 5-MHz bands while TD-SCDMA only requires one 1.6-MHz band. This feature is critical, particularly for densely populated economies like China and especially in metropolitan areas.

Second, the application of smart antennae can also enhance antennae gains because of their distinctive feature of adaptive beam-forming—the ability to target and track a specific object with a single signal beam rather than sending or receiving signals multidimensionally as with current GSM systems. This feature will significantly save energy and improve the power efficiency of the base station. Additionally, as smart antennae use an array of small power amplifiers, which are much cheaper than a single high-power amplifier, the cost of base stations can be reduced dramatically. The antenna array can also strengthen the reliability of the base station, as the system would remain working even if part of the eight sending/receiving systems were out of order.

Third, TD-SCDMA uses two asymmetric bands for uploading and downloading traffic, which are analogous in effect to the asymmetric digital subscriber line (ADSL) technology of a fixed line broadband system. The expectation with usage of the former is that 3G will be mainly used for downloading content from wireless portals.

Fourth, the application of software-defined radio enables the application of multiple features to be based mainly on software rather than hardware. This has several inherent advantages. On the one hand, it will reduce the size, weight, and cost of the system. And on the other, it should serve to overcome China's technical weakness in chip manufacturing technology.

The last and most important feature of TD-SCDMA is that it can enable a smooth transition from the current GSM system to 3G systems. TD-SCDMA is designed as a dual-band and dual-mode system. The 3G base stations can be installed in the same place as the GSM base station. In its coverage area, therefore, TD-SCDMA could support both GSM and 3G services. In areas that have not been covered by 3G base stations, GSM service would still be available. For this reason, the handset should be able to accommodate dual-band and dual-mode operation. With the application of software-defined radio, this should not prove difficult to achieve [21].

The TD-SCDMA system also has certain limitations. According to ZTE, a clear supporter of CDMA2000, TD-SCDMA is arguably more suitable to metropolitan areas than to remote areas [interview with ZTE, April 26, 2001]. As ZTE is now one of the major suppliers of Unicom Horizon's CDMA system, its viewpoint could be biased due to the fact that CDMA2000 will evolve much more easily from a CDMA or CDMA-1X network. In fact, CDMA2000 has raised concerns in China because it will depend upon the Global Positioning System (GPS) of the United States and thus security and reliability may be influenced by political factors outside China's direct influence.

Perhaps the true weakness of TD-SCDMA lies in the fact that it was proposed 2 years after CDMA2000 and WCDMA. The first test of the system was successfully conducted in Beijing on April 11, 2001. By that time, NEC, Ericsson, and other foreign vendors had already received a significant number of orders for the WCDMA system. In order to catch up, China currently needs to accelerate its development of TD-SCDMA. According to Datang, TD-SCDMA base stations will be available in mid 2002. The postponement of the WCDMA service launch by NTT DoCoMo (Japan) and Manx Telecom (Isle of Man) has provided Datang with very valuable time, which it can use to convert its current technology into commercial products in advance of the huge potential operator demand.

6.5   3G Licensing in China: Who Will Be the Winner?

According to Zhu [22], there are a small number of focused R&D teams working on 3G developments in China. Their purpose is to follow worldwide developments in 3G and to provide broad information and input to a 3G development strategy for China. The State Planning Committee is responsible for the coordination of these teams. The teams comprise the following:

- Datang's TD-SCDMA group, which is supported and sponsored by the MII;

- A research team under the auspices of the Ministry of Science and Technology focusing on fundamental research on WCDMA and CDMA2000;

- A group under the Science and Technology Commission of the Beijing municipal government, which is promoting LAS-CDMA (a technology to challenge future 4G systems);

- Groups working to individual domestic vendors, such as Datang, ZTE, and Julong, who control part of the intellectual property rights of a specific technology (e.g., ZTE with cdma2000).

The key question that now faces the Chinese government and operators is which 3G technology to adopt. As 3G in China has the potential to be a $100-billion market, any decision will have huge implications for operators, domestic manufacturers, and foreign vendors.

The MII has until now (June 2002) kept silent on which standard China will adopt. As a government department that is accountable for both telecommunication operations and IT manufacturing, it is difficult for the MII to take a technologically neutral stance, as overseas regulators, such as Oftel in the United Kingdom, have practiced in the field of 3G licensing. The MII has made little secret of the fact that it would like to see 3G technology being used as an impetus to repeat China's success in fixed line system manufacturing. Because of the fact that both China Mobile and China Unicom have been partly listed on overseas stock markets, the MII has been extremely reluctant to force operators to adopt a specific standard, as this might give investors the impression of over zealous government intervention. This would be unfavorable to the future listing of China Telecom and other Chinese telecommunications companies, placing China's position as a global telecommunications player in jeopardy. China's successful accession to the WTO also now constitutes a clear restraint on the visible expression and elaboration of the MII's interventionist ambitions.

Both China Mobile and China Unicom have shown enthusiasm for cooperating with Datang and have agreed to provide network capacity for the trial of the TD-SCDMA system. Li Mo-fang, Chief Engineer of China

Mobile, clearly indicated that China Mobile might use both TD-SCDMA and WCDMA for its future 3G network, as these two systems will enable a smooth upgrading of China Mobile's current GSM network to 3G. Additionally, the two technologies can supplement each other due to their different but complementary technical strengths [23]. For example, TD-SCDMA can be used in urban areas while WCDMA can be effectively deployed in rural counterparts. In an interview with ComputerWorld, Li Mo-fang further highlighted the point that "the real development of the mobile communications industry depends on the joint R&D efforts conducted by vendors and operators. If trials prove successful, and TD-SCDMA is proven to be a [robust] technology, [China Mobile] is going to adopt it" [24]. For Datang, this holds out the prospect of significant benefits.

Unicom Horizon's newly built CDMA network could provide an important future market for ZTE and North American vendors. ZTE is extremely confident about cdma2000 technology, for which it has been awarded important intellectual property rights. The encouraging progress that ZTE has made in CDMA will undoubtedly raise the Chinese government's expectations as to the role to be played by ZTE in future cdma2000 technology. ZTE's cdma2000 efforts can also reduce the threat of opportunity costs for the Chinese telecommunications industry as a whole in the eventuality that TD-SCDMA fails to provide satisfactory commercial products.

In recognition of the great potential significance of this huge market segment, most foreign vendors have undertaken strategic maneuvers in the past 2 years. In the beginning, many vendors lobbied heavily trying to convince the Chinese government that TD-SCDMA is not a feasible standard. Some directly asked the Chinese government "why China still wants to develop its own system? You are going to join the WTO, so just simply buy our products" [25]. The determined support of the Chinese government for TD-SCDMA, however, has clearly demonstrated to foreign vendors that rather than continue to question the axioms of Chinese policy directly, a more effective approach might lie in interfirm cooperation and partnership within enunciated technical parameters.

In fact, the last few years have played host to an accelerating partnership trend. Siemens was the first foreign vendor to partner with Datang for TD-SCDMA development. In July 1998, the two companies signed an agreement to jointly develop the TD-SCDMA system. By early 2001, Siemens had invested $1 billion in the project and established joint laboratories with Datang in Beijing, Munich, Berlin, Milan, Vienna, and London. More than 500 Siemens researchers are now working in China, and Siemens

3G mobile communication research center is now under construction in Beijing [26].

Hard on the heels of this agreement in September 2000, Motorola, Nortel, and Siemens, together with Datang, Huawei, China Mobile, and China Unicom, initiated the establishment of a TD-SCDMA Technology Forum, which aims to provide a platform for technology exchange among interested parties. In March 2001, the forum formally launched its Web site at http://www.tdscdma-forum.org. By the end of May 2001, there were 218 members, including foreign and domestic vendors, operators, chip producers, ICPs, research institutes, universities, and venture capital investors. Foreign vendors have evidently changed their attitudes towards TD-SCDMA. When doing business in China, the realization appears to have dawned that, instead of heedlessly attacking the rules of the Chinese "telecommunications game," market entrants need to adopt a policy of constructive engagement and adopt a stance of realistic pragmatism.

TD-SCDMA has certainly attained the status of a popular standard, but there is still some distance to travel before it is transformed into a commercial product. The standards issue surrounding the acceptability and dissemination of this product highlights important dilemmas in Chinese mobile telecommunications policy. Chinese decision-makers are naturally reluctant to squander investment in 3G development by encouraging operators to use only foreign products. At the same time, the question arises of whether they can afford to await 3G developments in other economic systems reactively before committing to an indigenous process of development. Until now, no mature 3G products have emerged in the global marketplace. China could thus be enticed into a relatively risk-free wait-and-see policy. The attendant risks of such a "risk-free" stance, however, could result in the country failing to gain the potential benefits unleashed by the commercialization of its own products. In this sense, critical issues surrounding 3G licensing in China may lie less in which allocation methodology should be used to issue licenses or define the costing of such licenses, but rather in the issue of the development and launch times to market mature TD-SCDMA products while attempting to effectively gauge the development and launch strategies of China's major international competitors.

6.6 Conclusion

3G licensing in China is an issue area with multiple policy dimensions. 3G systems are theoretically the engines of fundamental change in contemporary telecommunications systems. The economic implications of these technologies and their deployment are highly significant. Turning TD-SCDMA into a goose that lays golden eggs requires astute environmental scanning on the part of telecommunications strategists and a concerted response to the actions of competitors within tight time constraints. Effective joint efforts on the part of government agencies, vendors, and operators are certainly viewed within Chinese policy circles as critical to developmental ambitions.

Importantly, 3G licensing has also wrought breakthroughs in the policy framework of spectrum regulation. Although auctions are not the conventional weapon of choice of the Chinese regulator, the proposed method of pricing spectrum by benchmarking against prices prevalent in other countries while shifting payment from subscribers to operators should be viewed as signifying the beginning of spectrum commercialization according to market rules (again signifying a move from administrative to market governance). The increasingly internally competitive market coupled with scarce spectrum implies a more effective spectrum regulatory framework. The opening up of a tendering process for spectra used in the wireless local loop segment marks a positive step in a market direction.

The case of TD-SCDMA has also revealed that the Chinese telecommunications policy community has hardly been laggardly in response to 3G developments and, in effect, has busied itself examining policy lessons from afar. In fact, China appears to have taken a cautious and studied view of policy implications elsewhere in an effort to avoid their more dysfunctional consequences. Spectrum auctions are a case in point. Here as elsewhere, Chinese policy makers reveal a predilection for telecommunications policy particularly attuned to Chinese circumstances—telecommunications policy with Chinese characteristics. In the process, China's success with TD-SCDMA signals the country's intent to play a future proactive role on the international telecommunications stage.

References

[1]   Tan, Z. A., "Product Cycle Theory and the Telecommunications Industry—Foreign Direct Investment, Government Policy, and Indigenous Manufacturing in China," *Telecommunications Policy*, Vol. 26, No. 1, 2002, pp. 17–30.

[2]   Ministry of Information Industry, *The Development of the Chinese Telecommunications Industry: From 1949 to 1999*, Beijing, China: Ministry of Information Industry, 1999.

[3]   Ministry of Information Industry, *1999 Annual Report on China's Information Industry*, Beijing, China: Ministry of Information Industry, 2000.

[4] Yao, C. F., "Chinese Mobile Communications R&D Has Moved to a New Stage," *People's Posts and Telecommunications*, March 17, 2001.

[5] Xu, Y., "TD-SCDMA Is Critical to Chinese 3G Strategy," *Hong Kong Economic J.*, July 14, 2000.

[6] "China Unicom Launched Its CDMA Project," http://www.c114.net, February 17, 2000.

[7] "Chinese Military to Hand Mobile Business to China Unicom," http://www.total-tele.com, July 13, 2001.

[8] "WAP Users Surpass 8,000 in Beijing," *China Infonews* at http://www.cci.cn.net, November 18, 2000.

[9] NTT DoCoMo, http://www.nttdocomo.com.

[10] Xu, Y., "Return of the Tigers: Asian-Pacific Innovation in Mobile Communications," *Info*, Vol. 3, No. 3, 2001, pp. 195–206.

[11] 2001 Annual Report of China Mobile (HK), Hong Kong, China: China Mobile (HK), 2002.

[12] Lu, T. J., "The Development of Mobile Commerce in China," *Proc. Asia-Pacific Mobile Communications Symp.*, 2000, pp. 100–110.

[13] China InfoNet, "Wu Ji-chuan Talks About Future Development of the Chinese Telecommunications Market," http://www.c114.net, September 19, 2000.

[14] Article 28 of the Telecommunications Regulations of the People's Republic of China, Beijing, China: Ministry of Information Industry, 2000.

[15] Chan, J. C., "Regulation of the Telecom Resource," *Training Handbook of Telecommunications Regulations*, Beijing, China: Ministry of Information Industry, December 2000.

[16] Tender Invitation (No. GXTC-0104002-1) of the Ministry of Information Industry, June 29, 2001.

[17] Wang, F., "A Great Stride Towards 3G's Future," *Datang Group Newsletter*, April 1, 2001.

[18] Nan, M., "Chinese Standards," *Popular Computer Weekly*, July 10, 2000.

[19] Meng, W. S., and C. F. Yao, "The 'Golden Goose' of the Chinese Mobile Industry," *People's Posts and Telecommunications*, July 4, 2000.

[20] Li, S. H., "Radio Transmission Techniques in IMT-2000," *China Communications*, No. 37, November 2000, pp. 21–24.

[21] Yang, Y. G., "3G Evolution Strategy in the Chinese Context," *People's Daily*, December 20, 2000.

[22] Zhu, H. "The Impact of the Industrialisation Policy of Telecommunications, the Deregulation Process and Foreign Financing: China's Case," *Communications and Strategies*, Issue 41, 2001, pp. 37–61.

[23] "Chinese 3G Standard Might Be Named with 'T'," http://www.c114.net, December 5, 2000.

[24] Huang, G., "There Is Nothing Wrong with the Chinese 3G Standard, TD-SCDMA Is Not a Beautiful Mistake," *ComputerWorld*, March 27, 2001.

[25] Zhao, M., "3G: Uncertain Future of Telecom Operators," *China Economic Times*, December 21, 2001.

[26] "TD-SCDMA: The Future Is Not a Dream," *People's Posts and Telecommunications*, August 15, 2000.

7

Dancing with Wolves? WTO Accession and Its Impact on Telecommunications

The 14-year marathon negotiation between the Chinese government and key WTO member governments finally achieved a settlement. In December 2001, China formally joined the WTO. Thus, a new chapter opened for the Chinese telecommunications sector since a key part of the Chinese government's WTO commitments was a major concession over foreign direct investment in indigenous telecommunications operating entities.

7.1 Opening the Door: China's Commitment

According to the announced agreements, a foreign investment cap on basic telecommunication services operators was set at 49%, allowing the Chinese government to retain majority control. Foreign investors in mobile telecommunications were allowed a 25% share immediately upon China's accession to the WTO. This share will be allowed to rise to 49% after a 3-year period. For fixed network services, it will take 5 years to reach the investment cap. The arrangement further calls for the cap to be raised to 50% after 2 years for value added service and paging service. The foreign investment deal also applies to Internet service provision companies. Detailed Chinese commitments under its schedule of the General Agreement on Trade in Services are shown in Table 7.1.

Table 7.1
China's Commitment Under Its WTO Service Schedule

Type of Services	Percentage and Geographic Coverage of Foreign Investment Permitted					
	12/01–12/02	12/02–12/03	12/03–12/04	12/04–12/05	12/05–12/06	12/06–12/07
Basic telecom services—fixed	0%	0%	0%	25% in Beijing, Shanghai, and Guangzhou	35% in 17 cities	49% with no geographic restrictions
Basic telecom services—mobile	25% in Beijing, Shanghai, and Guangzhou	35% in 17 cities	No change	49% with no geographic restrictions	No change	No change
Value-added services and paging service	30% in Beijing, Shanghai, and Guangzhou	49% in 17 cities	50% with no geographic restrictions	No change	No change	No change

Source: WTO.

In accordance with China's commitments, the State Council promulgated its "Provisions on the Administration of Foreign-Invested Telecommunications Enterprises" on December 11, 2001. It provided criteria for foreign investors to set up joint ventures, such as minimum registered capital in the joint venture. In accordance with this provision, the procedures for applying for licenses were also clearly defined by the MII. A full version of this provision is provided in Appendix B.

The entry of China into the WTO has significant implications for the telecommunications industry in both China and other countries. For the former, this means an accelerated pace of transition for its telecommunications system, which has so far been faced with, and challenged by, only limited competition in selected services, into a market framework whose rules have been stipulated by multilateral parties, mainly developed and deregulated economies. Whether or not China can effect a smooth transition to a market-based system in this "great leap forward" (a question to be addressed in this chapter) remains a challenging issue.

The future prospects of the Chinese telecommunications industry in the post-WTO era have significant implications for other countries. For current WTO members who have committed to the Reference Paper and the Fourth Protocol of the WTO's General Agreement of Trade in Services (GATS, also called the Basic Telecommunications Agreement), the arrival of China indicates a significant diffusion of their shared values of open competition, fair rules, and effective enforcement into one of the world's largest telecommunications markets. Whether or not these values can be transplanted successfully into a telecommunications industry where a socialist ideology has been predominant for half a century remains an uncertain issue. The jury is still out on this question at the time of writing.

China's entry into the WTO also has significant implications for economies that remain outside the WTO's Basic Telecommunications Agreement. The WTO has highlighted the fact that the first group signatories to the WTO's Basic Telecommunications Agreement account for more than 90% of international telecommunications traffic and represent markets generating 95% of the $600 billion comprising global telecommunications revenues [1]. As pointed out by Oliver [2] and Tarjanne [3], however, the agreement only covers 19% of the world's population, and this mainly in developed economies. The 1.2 billion citizens of China have effectively doubled the size of the population covered by WTO agreements. If Chinese telecommunications operators turn out to be the immediate losers to their foreign rivals with WTO accession, it would be a serious discouragement for countries that so far have hesitated to commit on WTO obligations. Consequently, the further diffusion of the shared values of the developed economies into these developing economies would be frustrated, and the WTO's Basic Telecommunications Agreement would remain as accepted rules of the game for an exclusive "charmed circle" of players.

In fact, many countries have been concerned about committing to the WTO's Basic Telecommunications Agreement. Tarjanne [3] has provided a summary of major potential downside impacts of commitment to the WTO framework that have been concerning some WTO participants. In summary, these concerns include "the possibility of a decline in telecommunications revenue, an infringement of national sovereignty and a loss of control over basic telecommunication infrastructure."

These significant implications and concerns have strongly highlighted the importance of conducting an in-depth study of the potential impact of China's WTO membership on the development of its telecommunications. Questions needing to be addressed include the following: What are the motivators that have driven the Chinese government to undertake the decisive step of entering the WTO? What have been the key concerns in the telecommunications sector over China's accession to the WTO? Have the regulatory, legal, and organizational frameworks been reconfigured well enough to take full advantage of foreign direct investment?

A cache of literature exists on the manifold benefits resulting from foreign direct investment for host countries. Coughlin [4] found that foreign direct investment provided nearly 9% of all U.S. manufacturing jobs by 1988. He also concluded that foreign investment brings in capital and technology simultaneously. Blomstrom [5] claimed that foreign direct investment directly increases productivity by providing host countries with access to modern technology that they cannot provide themselves. Indirectly, inefficient local companies are forced to raise investment in physical and human capital due to increased competition after the entrance of foreign-owned enterprises. Other benefits occurring in host countries include increased competition, reduced cost of capital, and positive externalities in technology and management [6].

In addition to the above advantages, studies have also relieved some concerns over the negative impact of foreign direct investment. Again, Coughlin [4] found that concerns over the "headquarter effects," or the extent to which foreign owners will use the host country simply for production and shift R&D activities outside its domain, is, in fact, overstated. His research has indicated that spending by foreign investors per capita on R&D in host countries is almost the same as that in their own countries. Ondrich and Wasylenko [7] have claimed that local employees are not disadvantaged by foreign employers in terms of wage and salary payments, and some of these employees are, in fact, paid more handsomely than their counterparts in domestic companies. Graham and Krugman [8] insist that foreign direct investment is not a major contributor to the trade deficit, as a certain portion of the products from these foreign invested companies are for export rather than for seeding markets in the host countries. The host country should also take into account other benefits from foreign direct investment, such as possible technology transference.

The above studies positively favor foreign direct investment. Most of them, however, were based on cases in developed economies rather than developing ones. Nor are they directly relevant to the issue of concerns over WTO entry. Specifically, literature on the impact of foreign direct investment in the telecommunications industry is limited because of the fact that such investment has been strictly regulated until recently in most countries, even in the most liberalized economies, such as the European Union and the United States. Arguably, the first wave of multilateral foreign direct investment in telecommunications formally began in February 1998, when the WTO's Basic Telecommunications Agreement came into force with individual parties committed to different degrees of openness [1]. Continuing reflection on and analysis of the impact of foreign direct investment on telecommunications is of importance to both developing and developed economies.

This present chapter takes into account broad political, economic, and financial contexts and offers an up-to-date review of China's policy stance on foreign investment in its telecommunications sector. It reveals the fact that the Chinese government is, essentially, adopting a middle approach by taking advantage of foreign investment on the one hand and securing firm control over the telecommunications operation on the other. Controversies that could arise in the telecommunications sector after China's accession to the WTO are also analyzed here. The chapter suggests that current Chinese telecommunications legal, regulatory, and organizational frameworks need to be effectively restructured, while privatization is critical for avoiding potential conflicts of interest in the post WTO era.

7.2 Joining the WTO: A Strategic Step by the Chinese Government

Recent history reveals strongly that the Chinese government has displayed great enthusiasm for joining the WTO, despite the fact that there exist concerns over foreign direct investment in telecommunications and other industrial sectors. Both political and economic considerations have led to the adoption of this policy stance. Politically, China has preferred to join a multilateral forum like the WTO to iron out potential disputes with trade partners rather than adopting a bilateral negotiation approach, as the latter is usually vulnerable to political intervention. For example, China has had to fight a determined and protracted battle to obtain most-favored-nation (MFN) trade status from the United States on an annually renewable basis. Each year, the U.S. Congress conducts an intense debate over such issues as the human rights record of the Chinese government. Although MFN status has always been granted, thanks to intensive lobbying from the U.S. government and industry, it has been productive of tensions and uncertainty in trading relationships between the two countries. Eliminating such uncertainty is critical for China as imports and exports have played a rapidly increasing role in China's economy. For example, the share of imported and exported goods and services in

China's GDP has increased from 35.1% in 1990 to 48.9% in 1999 [9]. By joining the WTO, China should in theory be able to solve conflicts over trade with individual countries under the commonly agreed prin- ciples of the WTO and obtain permanent normal trade relations status from its major trade partners. In fact, in December 2001, only 2 weeks after China's WTO membership, the United States granted China permanent status as a "normal" trading partner with immediate effect from January 1, 2002.

Another key political consideration relates to the issue of the unifica- tion of China [10]. After the handover of Hong Kong and Macao, Taiwan now remains as the sole territory with which the Chinese government is committed to integrate. Due to reluctance on the part of the Taiwanese administration, there has been no direct trade in posts, telecommunications, and transportation between mainland China and Taiwan. All mail, telecom- munication traffic, passengers, and goods have to transit via third-party ter- ritories, primarily Hong Kong and Macao. After becoming members of the WTO, both mainland China and Taiwan have to commit on obligations for free trade whereby direct trade of posts, telecommunications, and transpor- tation can appropriately be achieved. Increased interdependence would positively facilitate the future peaceful unification of China.

In addition to the above political considerations, the determination of the Chinese government to transform China's highly centralized and planned economy into a socialist market economy is perhaps the most pow- erful driving force behind China's desperate bid to enter the WTO. Chap- ters 2 and 3 clearly revealed that the Chinese government has taken a strong and consistent stance in introducing market economy mechanisms into the telecommunications sector, but in a measured step-by-step approach. Such reform schemes as the Contractual Responsibility System have effectively decentralized the once highly centralized system of power and made consid- erable strides in turning the operation of telecommunications from a politi- cal instrument into a profit-driven business. The licensing of China Unicom marked the second round of reform, and the preliminary success of deregu- lation in the cellular market has strongly encouraged the government to move even further ahead. Consistent with developments in telecommunica- tions, the government also intends to liberalize other industrial sectors. From this perspective, it is not unreasonable to infer that joining the WTO was not a reluctantly adopted policy option, but rather a proactive strategic step in the Chinese government's well-laid scheme of evolutionary "policy reinvention." For telecommunications, this marks the beginning of a third round of reform.

7.3   Foreign Investors: Wolves in the Backyard?

The Chinese government has historically taken a conservative stance over foreign direct investment in telecommunications in a context in which China's sovereignty was viewed as vulnerable to outside challenge. Unpleasant sovereignty disputes involving confrontation with foreign operators, such as the Danish GNTC, as reviewed in Chapter 2, established unfortunate historical precedents. These cast a long shadow forward to greatly influence the future policy stance of the Chinese government. The result has been that clear priority has continued to be accorded to the issue of sovereignty in the formulation of telecommunications policy.

Another clear concern over foreign direct investment is the question of whether it is still too early to expose Chinese domestic operators to intensified competition after foreign operators move into the market. For a country with a teledensity of 13.9 mainlines per 100 inhabitants (December 2001), which is similar to the level of the United States in the 1920s and much less than the 46% when the Bell System was split up in 1984, the insistent question has arisen as to whether it is premature for China to open its telecommunications market for foreign direct investment. Officials from the telecommunications sector have continually warned central government that immediate market opening to foreign operators is akin to inviting wolves into the domestic backyard. According to such arguments, the newly developed telecommunications industry in China will turn out to be a hapless victim of aggressive and greedy invaders. In the light of such fears, the Minister of the Information Industry reportedly submitted a resignation letter to the State Council protesting against the Premier's compromise on foreign direct investment in China's telecommunications sector during his negotiations with the U.S. trade representative on China's entry into the WTO [10].

In addition to the above concerns, a continuing anxiety is that national security might be threatened by foreign direct investment. That is, foreign ownership of telecommunications facilities might enable state secrets to all too easily leak out of China. Additionally, the telecommunications infrastructure might escape national control during a state emergency, such as a war [11]. The latest evidence of such concerns is the newly announced "Regulation on Security Control over Computer Systems and the Internet" by the MII that goes so far as to forbid organizations from purchasing security-protection software from foreign vendors [12].

In fact, experience in other economies indicates that these concerns have been overly exaggerated. For example, as to concerns over national sovereignty and loss of control over basic telecommunications infrastructure,

Tarjanne [3] demonstrates that the GATS of the WTO is structured to allow for the primacy of governments' decisions on the pace of liberalization. He argues, "participants can commit to different levels of market opening, depending on the stage of development of their telecommunications sector, their current state of sectoral reform, and the government's assessment of its national interests." In fact, as the WTO adopts a one-member-one-vote policy, weak economies can use the WTO as a forum to defend their interests more effectively. For example, when Thailand joined the WTO (then GATT) in 1982, it was able to defend its interests successfully over bilateral trade on rice and tobacco with the United States, which had been impossible before then [13]. Joining the WTO is, in fact, an effective way of enhancing, rather than weakening, national sovereignty.

The second major concern regards the timing of WTO accession, due to the current low telephone penetration rates in China. It is true that the penetration rate is much lower than the penetration rate in the United States when its telecommunications market was liberalized in 1984, but by 1994 all the signs were in place suggesting that the market was being transformed from a sellers' to a buyers' variant [14]. Given this market reconfiguration, the Chinese telecommunications industry is currently facing similar challenges to those faced by early-mover countries in the early 1980s, namely pressures for improving efficiency and triggering more demand. As foreign evidence has shown, these objectives can only be achieved by strengthening market competition. A study by the International Telecommunications Union has revealed that the total volume of international direct dialing (IDD) traffic has increased steadily in countries that have liberalized the telecommunications market, while the opposite situation has obtained in countries that retain monopoly operation [15]. For example, in Chile, a "30% reduction in prices for international calls from Chile to the United States (due to competition) led to a 260% increase in traffic volume on that route" [2]. This implies that there is immediate and insistent need to open the current Chinese telecommunications system to more fierce competition.

Lastly, concerns that national security will be threatened by foreign direct investment are probably greatly exaggerated. Early experience in the United States clearly indicates that ownership of telecommunications infrastructure is not a foolproof method of safeguarding state secrets. In Section 310(b) of the Telecommunications Act of 1934, foreign ownership and operation of American radio systems were strictly controlled. Ironically, in 1941, Imperial Japanese diplomats readily transmitted encoded messages relevant to the imminent attack on Pearl Harbor from the United States to Tokyo over facilities operated by American-owned telegraph carriers. Thus, "even in the relatively simple technological era in which Congress enacted section 310(b), foreign ownership restrictions failed to protect America's national security" [6]. In fact, the most effective way to protect national secrets is to properly protect the information resource rather than the transmission instrument, as the information can be delivered through any kind of transmission medium. Additionally, such specific arrangements as the "golden share" and legal safeguards, such as Section 606 of the above mentioned U.S. Telecommunications Act, which grants power to the president to take over the telecommunications system in cases of emergency, like a war, might be more effective than blanket prohibitions forbidding foreign direct investment.

In brief, experience in other economies has clearly shown that most of the concerns over entering the WTO have been overly exaggerated. It will take time, however, for the Chinese government to relieve itself of such concerns fully. Until December 2001, that government had successfully blocked foreign direct investment in the telecommunications sector. In a regulation promulgated by the former MPT, it was clearly specified that foreign direct investment in the telecommunications market was strictly banned [16]:

> No organizations and individuals outside China or solely foreign-funded enterprises, Sino-foreign joint ventures and cooperative businesses on the territory of China shall invest in, operate or participate in the operation of telecommunications services in China.

As members of a capital-intensive industry, however, all Chinese telecommunication operators have potentially insatiable needs for financial investment. They have been quick to recognize the urgency of obtaining such foreign investment. In past years, they have been forced to find methods of bypassing government regulations that impose strict controls over foreign direct investment in the indigenous telecommunications sector and challenged the "forbidden zone." As a result, policy over foreign investment has been gradually modified in recent years.

7.4 Financial Urgency for Foreign Direct Investment in Chinese Telecommunications

In order to understand the evolution of China's policy on foreign investment in the telecommunications sector, it is necessary to provide a comprehensive review of how China's telecommunications system has been funded historically. In fact, the funding development of telecommunications investment in China can be divided into four stages: (1) from 1949 to 1981, (2) from 1982 to 1993, (3) from 1994 to 2000, and (4) 2001 and onwards.

7.4.1   Stage 1—1949 to 1981

During this period of time, all investment in telecommunications emanated mainly from the government together with some of the enterprises' own funds (including retained profits, depreciation, and overhaul). Given the limited scope of financing channels, the entire investment in telecommunications for these 32 years was just 6.4 billion yuan (around $3.7 billion according to the 1981 exchange rate) [17].

7.4.2   Stage 2—1982 to 1993

In 1982, the government designated the telecommunications sector as one having strategic economic significance and granted the MPT the following preferential treatment by exploring new resource avenues and incentives:

1. "Three 90 percents," installation fees, and surcharges, as reviewed in Chapter 2;
2. Priority to be given to the MPT regarding infrastructure capacity quotas, allocations on foreign currency importation and foreign government loans;
3. Government facilitation of MPT upgrading of telecommunication facilities and land confiscation. Waivers would also be given to the MPT on fees payable for toll roads and bridges.

Such treatment had fostered the development of the MPT and overall investment over this period reached 90.5 billion yuan ($15.7 billion according to the 1993 exchange rate). This represented 14 times the level of the previous 32 years' development. Financing channels during this period were flexible and diversified. The main sources of funding came mainly from internal funds, installation fees, surcharges, and government investment. Debt financing was mainly obtained from soft loans and supranational credits, such as those from the World Bank and the Asia Development Bank.

7.4.3   Stage 3—1994 to 2000

During this period, the development of telecommunications and its funding in China underwent drastic changes:

1. With considerable changes in taxation, fiscal, and monetary regulations, the whole Chinese economy increasingly migrated towards a market structure.

2. The monopoly operation of the MPT was broken and China Unicom was established to compete with China Telecom.

3. As the telecom sector became economically viable in China, soft loans from foreign governments and international organizations began to shrink. Export credits with suppliers' subsidies (with a grant element of 20% of the contract value) were adopted in 1994 and 1995.

4. Along with diminution of external funding went diminishing regional investment. Free government appropriation was shifted to loans with market interest rates and again was diminishing. It was becoming increasingly difficult to levy surcharges on customers, and installation fees were decreasing.

5. The internal funding mix changed. As the telecommunications sector had now embarked on a new development path, the government abolished the "three 90 percents" preferential policy. Instead, it allowed telecommunications enterprises to adopt a double residual value decline in depreciation policy so as to facilitate enterprises in claiming faster depreciation, building up a bigger network renovation pool and paying less profits tax.

Faced with this increasingly tough environment in financing telecommunications development, both China Telecom and China Unicom were obliged to fashion new financial strategies. One such strategy used by each of these operators has been the utilization of deferred payments from vendors and a policy of facilities leasing. This method gained popularity between 1996 and 1998 but was far from sufficient in facilitating the rapid expansion of the network. A breakthrough was required to extend the funding pool.

One strategy adopted by China Telecom was the opening of a branch in Hong Kong—China Telecom (HK)—in September 1997, which was redesignated as China Mobile (HK) in 2000 following the divestiture of China Mobile from China Telecom. Its main function is to absorb foreign investment, and reinvest it in the Chinese cellular market. The operation of these networks is not controlled by China Mobile (HK) but by China Mobile's provincial companies, so the reinvestment from China Mobile (HK) is not considered as foreign direct investment but foreign indirect investment. In this way, China Mobile (HK) has taken over 21 provincial cellular networks.

China Unicom's preferred strategy in previous years (as highlighted in Chapter 4) was the so-called CCF joint investment scheme. According to this scheme, a company owned by local government or one of China Unicom's shareholders establishes a joint venture with foreign companies. The objective of this joint venture is similar to the strategy of China Mobile (HK)—that is, to attract necessary investment funds. It then reinvests them in the construction of mobile communications infrastructure, leaving China Unicom responsible for operational management. Cash flows are shared between the joint venture and China Unicom according to an agreed-upon ratio. In this way, China Unicom has, indeed, bypassed government's restrictions on foreign direct ownership [18]. The disadvantage of this scheme, however, is that China Unicom will never be master of its own fate and will permanently remain as a contracted agency for operating networks without owning the network itself. Ownership of the networks remains firmly in the hands of investors in the joint venture.

In late 1998, the MII conducted a review of China Unicom's CCF investment scheme and had it terminated. In return for this, China Unicom received a generous loan totaling 5 billion yuan ($604.8 million) from the Bank of China. In June 2000, following the financial exemplar of China Mobile (HK), China Unicom successfully completed its initial public offering in stock exchanges in Hong Kong and New York and raised a total of $5.65 billion via its branch company China Unicom (HK). Cellular networks from 12 Chinese provinces were listed.

7.4.4 Stage 4—2001 Onward

The telecommunications industry in China is now facing extremely difficult financing problems. If China were to raise its teledensity from 20.1 per 100 inhabitants by the end of 2000 to 50% by the middle of this century on the assumption that population growth would increase from current levels of between 1.2 billion to 1.6 billion, then total investment would be 3.5 trillion yuan ($422.7 billion) [19]. If consideration is also given to Internet investment, 3G mobile communications, and other potential innovative services, then the total volume of investment required in telecommunications would be doubled, or even quadrupled.

Simultaneously, orthodox financial resources, as utilized in the stage from 1994 to 1999, are continuously shrinking. The installation fee, which has counted for more than 20% of total investment in past years, has been formally abandoned since July 2001 as a result of growing pressure from the public. Additionally, the erosion of the current international settlement system—a system used to compensate domestic operators for terminating IDD calls by their foreign operator counterparts and which has contributed billions in hard currency to China Telecom in past decades—will also tighten future budgets for all Chinese domestic telecommunications operators.

To solve these problems, the most effective solution is undoubtedly that of attracting foreign investment. The experience of China Mobile (HK) and China Unicom (HK) has demonstrated that foreign capital can be the most effective resource for financing the Chinese telecommunications industry. So far, the foreign indirect investment schemes of these two companies, which have involved the establishment of a branch in Hong Kong and listing selected provincial networks on overseas stock markets, appear to be the sole favored way of attracting foreign investment by the Chinese telecommunications sector. Such schemes do not challenge existing state control over telecommunications operations—foreign investors have acquired an ownership stake in the two Chinese mobile networks indirectly via their listed branches in Hong Kong, but have been effectively prevented from participating in day-to-day operations.

Foreign indirect investment schemes of China Mobile (HK) and China Unicom (HK), however, are not necessarily the optimum choice for Chinese telecommunications operators. Coughlin [4] found that such externalities as advanced technology and experience can be transferred to the host country simultaneously with the invested capital in the direct investment model, while the above foreign indirect investment schemes run by China Mobile (HK) and China Unicom (HK), in fact, act as filters that allow in capital flows but block the transference of these externalities. As a result, Chinese operators have only benefited from increased capital, but not improved efficiency. In the meantime, as argued by Ginn [20] in his study of American foreign investment, such indirect investment increases the financial risk for investors, as they cannot participate in and monitor the operation of the business effectively. In this case, it remains doubtful if foreign investors will retain their loyalty to the current investment schemes of China Mobile (HK) and China Unicom (HK). This is especially true at a time of increasingly evident global deregulation, which has generated tremendous investment opportunities elsewhere in the world, such as in India [21].

The above review of telecommunications investment in China clearly indicates that financing the future expansion of telecommunications has become a growing problem facing the Chinese telecommunications industry. Foreign capital is undoubtedly the most effective resource for meeting the dramatically increasing demand. For many overseas investors, however, China Mobile (HK)'s currently preferred foreign indirect investment scheme is at most a second-best choice. To sustain its impressive growth in telecommunications, China has to break its decades-long restriction on foreign direct investment. Joining the WTO and directly attracting foreign investment is clearly a critical step for the future development of Chinese telecommunications. No other viable alternatives appear to be available.

7.5 Policy Position on Foreign Investment: A Middle-Way Approach

The analysis presented above clearly demonstrates that there is political, economic, and financial urgency in China's telecommunications sector embracing the WTO. Simultaneously, as reviewed earlier, there remain great concerns over foreign direct investment in Chinese telecommunications, although these concerns are arguably overexaggerated.

Figure 7.1 outlines those factors that have influenced the Chinese government's policy stance on foreign investment in the telecommunications sector within the specific context of China's accession to the WTO [22]. Factors on the left side are depicted as those that favor China's accession to the WTO and which open the door for foreign direct investment. Factors on the right side, however, raise concerns over foreign direct investment in telecommunications. The Chinese government is thus caught in a dilemma: on the one hand, it favors a well-financed competitive telecommunications market; on the other, it intends to maintain and secure its control over the telecommunications system. In such a case, the preferred strategy of the Chinese government has been to take advantage of foreign investment while, at the same time, retaining its control over telecommunications operations. This is one of the reasons why the foreign indirect investment schemes of China Mobile (HK) and China Unicom (HK) have been favored by the government. It also constitutes one of the main reasons for the government spending so many years and such effort in defending the foreign direct investment cap of 49% for basic services and 50% for value-added services in its WTO negotiations. Grappling with the contradictions of recognizing the imperative of attracting in foreign capital while avoiding the perceived downside of increased foreign influence has led many commentators to view the Chinese government as schizophrenic. As one commentator on the Chinese telecommunications scene has noted, the Chinese leadership elite "must simultaneously lure in foreigners and hold them at bay" [23].

A clear understanding of the Chinese government's policy stance on foreign direct investment in the telecommunications sector is of great

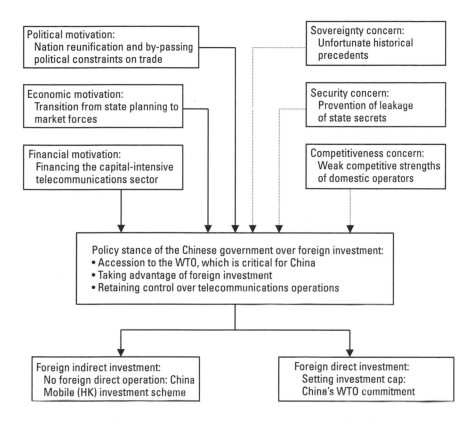

Figure 7.1 Impact of external factors on Chinese government's policy stance regarding foreign investment in telecommunications operation.

importance for overseas investors when formulating investment strategy. Recently, the Hong Kong government proposed the initiative of setting up a Closer Economic Partnership Arrangement (CEPA) with Mainland China, intended to confer early-mover advantages on Hong Kong companies in dealings with China proper. In a recent focus group discussion on CEPA hosted by the Hong Kong General Chamber of Commerce, representatives from selected Hong Kong telecommunication operators suggested that the Chinese government should grant management control to Hong Kong operators in lieu of full ownership control restricted by the investment cap of 49% and 50%. In fact, no such terms on management control exist in the newly published "Provisions on the Administration of Foreign-Invested Telecommunications Enterprises." Clearly, the issue of management control deeply concerns the Chinese government. It has defended strongly the investment cap in its WTO negotiations and will not readily concede management control as an ownership proxy. Requests for management control, even from the Special Administrative Region of Hong Kong, are unrealistic, and the Chinese government will undoubtedly continue to reject such overtures.

Nevertheless, China is now a formal member of the WTO and the door will be gradually opened for foreign investors according to the committed timetable of the Chinese government. For the government, the contemporary challenge lies less in attempting to block foreign direct investment than in exploiting the full benefits of such investment. Demolishing the "forbidden zone" of foreign direct investment constitutes a revolution that involves drastic restructuring of regulatory, legal, and organizational frameworks.

7.6   Challenges in the Post-WTO Era: Is China Well Prepared?

In fact, in order to prepare for the potential challenges immanent in the post-WTO era, China has already taken several drastic preemptive steps to restructure its telecommunications industry. These have included the establishment of the MII, the separation of telecommunications and postal service operations, and the splitting up of China Telecom. Unfortunately, as will be shown below, some reform schemes have been subject to certain limitations, which could imply the emergence of difficult problems in China's post-WTO era. This section provides an in-depth analysis of these reform schemes and highlights key factors that should be considered in any future scenario of telecommunications reform.

7.6.1   Regulatory Framework

According to the State Council, one of the MII's central commitments lies in the coordination of all telecommunications networks, public and private. However, as all the private networks continue at present under the ownership of individual ministries, the MII, in fact, is very weakly located for effective coordination. An illustrative example is the coordination of facilities colocation between the telecommunications and cable networks.

The cable network, owned by the former Ministry of Radio, Film and Television and currently by SARFT, has expanded extremely rapidly in the last decade. The annual growth rate of cable network subscriptions was 36.8% from 1992 to 1996 [11]. According to SARFT statistics,

84.76 million households, or 24.3% of households, had subscribed to the cable network by August 2001. This vast coverage of the cable network and the availability of such technologies as cable modems strongly motivated the cable network operator to provide services beyond the staple offerings of traditional TV broadcasting. It set up Internet and data broadcasting branches and has strongly challenged China Telecom's monopoly status over telecommunications services. Even after the former Ministry of Radio, Film and Television was downgraded to the State Administration of Radio, Film and Television in 1998, the sensitive status of political propaganda in China still confers upon the SARFT strong bargaining powers to defend the cable network's interests. In December 2001, the China Radio, Film and Television Group was formally established. One of its strategies is to integrate currently segregated broadcasting networks into a nationwide network including both cable and satellite facilities.

In reaction, China Telecom has tried hard to block the further expansion of the cable network. According to the MII, China Telecom should share its facility resources with the cable company on mutually beneficial terms. In reality, however, it is hard to achieve such an agreement because of the absence of clear guidelines and the weak power position of the MII. Conflicts between the two parties have become increasingly fierce. In Shenzhen, for example, whenever the cable operator inserts its cable into a duct belonging to China Telecom, the latter attempts to extract them. In November 1999, such conflict of interest nearly erupted into armed combat. The cable operator called for help from the Public Security Bureau when its cable was pulled out after several previous attempts, while the Shenda Telephone Company, a local branch of China Telecom, solicited the assistance of armed policemen. Soldiers on each side sat inside vans with weapons in hand, while representatives from each company held on-site negotiations. The cable company accused the Shenda Company of attempting to destroy the "propaganda throat" of the Chinese Communist Party, while, in turn, the Shenda Telephone Company claimed that they were merely protecting the safety of national telecommunications assets, a duty conferred upon China Telecom by the State Council as a component element of the company's portfolio of commitments. Finally, through the intervention of the mayor, the conflict was terminated [unattributed interview, December 5, 1999].

One way to avoid such conflicts in the future might lie in merging the current SARFT with the MII. One of the emergent advantages of such a rebrigading would be a reduction in the political bargaining power of the cable operator once subjected to MII direct regulation. Another solution could lie in enhanced coordination between cable and telecommunications networks. The strengthening of the MII's coordinative and regulatory authority is critical for future foreign investment, as a chaotic regulatory playing field contains the threat of increasing investment risk while simultaneously decreasing incentives to foreign investors with an inevitable hemorrhaging of confidence.

Another advantage is that regulation over content and transmission could be bettered coordinated if these two authorities are merged, and a protracted process of licensing converged services, such as occurred when Hong Kong Telecom attempted to obtain permission for providing Interactive TV service, could be avoided. When Hong Kong Telecom (HKT) applied for a license for Interactive TV service in 1996, the Hong Kong Cable Company warned the government that issuing a license to HKT would infringe its exclusive franchise over cable TV broadcasting. As two government departments separately regulated telecommunications and broadcasting, it took a considerable time to achieve the necessary compromises to end the controversy. In 1998, the Information Technology and Broadcasting Bureau was established to coordinate all relevant issues regarding broadcasting, information technology, and telecommunications [18].

The latest initiative of the Chinese government to coordinate the development of these networks was the establishment of the State Informationization Leading Group in late 2001. The group is headed directly by Premier Zhu Rong-ji. Its major function is to coordinate the development of all networks. The poor performance, however, of the former State Joint Conference on National Economic Informationization and the State Council Steering Committee of National Information Infrastructure during the period of 1994 to 1998 (as reviewed in Chapter 5) casts doubts on the likely effectiveness of this newly established group.

7.6.2 Industrial Reorganization

Since the establishment of the MII in 1998, China has undertaken three steps on the road of industrial reorganization. The first step, as noted, lay in separating postal and telecommunications service operations. Here the central purpose was to render the subsidy to postal services more transparent, so that both services could operate on the basis of real, fully attributable costs. This was a positive step for both services: for telecommunications, a heavy burden has been alleviated; for the postal services, a decreased subsidy undoubtedly constitutes an important incentive for efficiency improvements. Following separation, the revenue growth rate of the postal service reached 28.8% in 1999, the highest in recent years, while its operational loss was reduced by 11.26 billion yuan ($1.3 billion) [24].

The second step taken in organizational reform was to split China Telecom into four independent companies in 1999 according to four categories of service: China Mobile, China Satellite, Guo Xin Paging, and China Telecom (Guo Xin Paging was later merged with China Unicom). The monopoly status of China Telecom remains well protected because of its ownership of the vertically integrated fixed network, while China Mobile is left to compete with China Unicom over mobile phone services.

Liu Cai, former director of the Department of Policy and Regulation of the MII, provided his interpretation of the rationales behind this restructuring in a MII workshop in April 2001. According to Liu, the government had conducted investigations on the use of different industry structural models throughout the world. Adopting animalist allusions, he characterized the seven Bell operating companies in the U.S. market as seven "small tigers" while the European phenomenon of a giant incumbents continuing to compete with smaller new entrants could be seen as an "elephant playing with several monkeys." The European model was clearly preferred given China's ambition to build China Telecom into a national flagship system to compete in the international market. Adoption of the U.S. model could compromise such ambitions by reducing competitive effectiveness through reduced company size and the increased cost of interconnecting individual networks.

Another rationale for splitting China Telecom according to services rather than regions was that telecommunications services are becoming increasingly nationalized and globalized. Separation of companies according to geographic regions might be incongruent with developing technological trends [25]. Theoretically, this might be true. For regulators, however, what is of even greater significance is the facilitation of effective competition. The vertically integrated operation of long-distance and access networks by China Telecom will not leave much space for the maturation of new fixed network operators, and facility-based competition could, in consequence, remain as a remote objective.

In December 2001, the State Council announced its third step in industrial restructuring. China Telecom is currently being split into competing northern and southern firms. According to the State Council, China's telecommunications restructuring will adopt a "5 + 1" solution. China Netcom and Jitong will merge to "roost" in the northern 10 provinces of China Telecom as a single operator renamed as China Netcom, while China Telecom will preside over the remaining 21 southern provinces. This leaves China Mobile, China Unicom, and China Railcom operating independently with China Satcom as the "plus one." China Satcom is a new satellite company that was established in December 2001. It was formed through a merger of several individual satellite companies: China Telecommunications Broadcasting Satellite Corporation, China Orient Telecom Satellite Company, China Space Mobile Satellite Telecommunications Company, the Hong Kong–based ChinaSat Corporation, and China Post and Telecommunications Translation Service, a translation service company. On May 16, 2002, China Telecom and China Netcom were officially relaunched. Table 7.2 illustrates the market structure for basic telecommunications services after the separation of China Telecom.

Although the U.S. model was partly incorporated into this round of restructuring, this latest reorganization appears disappointingly devoid of strategic intelligence as little progress appears to have been made in the facilitation of local fixed network competition. The only two winners seem to be China Netcom and Jitong Telecom who have received a windfall of local networks in 10 provinces. Currently, the two giant tigers, China Telecom and China Netcom, remain as vertically integrated dominants in southern and northern China. More hard work apparently remains to be done to craft a strategy of facilitating future competition over fixed networks.

7.6.3   Ownership

Because of obvious ideological concerns, the entire current constellation of telecommunications service providers (China Telecom, China Mobile, China Unicom, China Railcom, China Satcom, and China Netcom) are all majority-owned by the state. The system of state ownership has led to a problematic government policy stance [18]. The government clearly harbored the ambition of achieving the benefits of competition on the one hand while being reluctant to invest in duplicative networks on the other.

Table 7.2
The Chinese Telecommunications Market Structure After Restructuring

	Fixed Local	Fixed Long-Distance & IDD	Mobile Cellular	Satellite Transmission
China Telecom	*	*		
China Unicom	*	*	*	
China Mobile			*	
China Netcom	*	*		
China Railcom	*	*		
ChinaSat				*

Additionally, to enhance the value of the state assets, the regulator has tended to protect the interests of the state rather than that of the public. Evidence in support of this argument emerged when China Mobile and China Unicoms' Shandong provincial branches declared a price war in September 1999. This attracted 380,000 new GSM subscribers in a period of 20 days. The MII immediately intervened, emphasizing that the tariff for cellular service should be decided by the regulator and should not be set without permission from the MII. The rationale adopted fell under the banner of the prevention of overcompetition and subsequent loss of state assets [26].

This kind of protection might be seen as reasonable in a situation of unambiguous state ownership. However, in the situation (as at present) when state-owned operators have begun the process of establishing partnerships with foreign investors, there is growing suspicion that the regulator will regulate in favor of the "particular interests" of those investors rather than in favor of the general public interest. If regulatory practice sets out to protect investor interests through the suppression of successive waves of competition, the main victim will be the Chinese consumer. Such an outcome would inevitably appear to contradict the spirit and intent of WTO membership. In this scenario, the privatization of telecommunications operations acquires urgency, so that the regulator can concentrate on the primary strategic priority of enhancing competition.

Another advantage of privatization lies in the complete removal of barriers against future merger and alliance activity between Chinese telecommunications operators and their foreign partners. According to Shi Wei of the State Council, in order to strengthen Chinese operators' competitive strengths, China Telecom should be encouraged to establish a major alliance with key operators in Asia, and this alliance should become a leading operator, dominating this region [27]. This provides a clear signal that the Chinese government has clearly realized that the WTO also provides Chinese operators with a golden opportunity to penetrate foreign markets, although it is questionable why the alliance would be limited only to Asia at a time when telecommunications is increasingly globalized. Nevertheless, no matter which partners Chinese operators are targeting, as merger negotiations in 2000 between Singtel and Cable & Wireless HKT have shown, state-owned status will place them in a vulnerable bargaining position.

Singtel is the dominant telecom operator in Singapore, and the government holds 78% of shares indirectly via its investment arm—Temasek Holdings. In January 2000, Cable & Wireless announced that it was negotiating with Singtel on the possibility of transferring the ownership of Cable & Wireless HKT to Singtel. Hong Kong was alerted to this deal and the legislator, Chungkai Sin, pointed out that this merger might enable the Singapore government to intervene in Hong Kong's domestic telecommunications market and usurp the Special Administrative Region's position as the information hub of Asia [28]. Finally, a local company named Pacific Century CyberWorks entered a counter-bid to take over Cable & Wireless HKT [29]. A parallel case was that involving the German giant, Deutsche Telekom. When this mainly government-owned company attempted to take over Telecom Italia, it was the local company Olivetti acting as a white knight that eventually succeeded as a result of intervention from the government. The central lesson of both cases was the fear that a foreign company, largely state-owned or -influenced, could take over a country's national telecommunications system and render it subject to outside control. Fears (however overblown) emerging from such experiences have had an impact on emerging thinking in Chinese telecommunications circles. Privatization is currently emerging as less a threat to the integrity of the Chinese tele- communications system than a potential protective ally. Arguably, privatization may prove a critical factor in paving the way for Chinese operators to enter into future merger and alliance formation with foreign operators. A strong Chinese presence in such formations, providing China with the resources to play in the international arena (and achieve a degree of foreign market penetration), may constitute useful buffers to any foreign-state-dominated predators with ambitions to dominate the Chinese telecommunications marketplace.

In fact, the current industrial structure of the Chinese telecommunications market is very well-prepared for privatization. In several countries where privatization was deployed before the market was liberalized, the growth of telecommunications has slowed down. In Mexico and Venezuela, for example, a monopoly franchise of 5 years was granted to a small group of investors from the private sector. As a result, both countries have suffered from a period of declining growth, as a wider circle of investors has been prevented from market entry until the expiry of the conferred monopoly franchise (Figure 7.2). Monopoly operation has, in this case, achieved the short run gain of monopoly rents for the few while sacrificing long-term profitability and growth. In China, however, where most market sectors have been liberalized, such a denouement has effectively been prevented. Investment entry barriers have already been breached with the prospect of fierce competition and rapid growth subject to a multiplier effect with eventual privatization.

The early spring seeding of such privatization may, in fact, already have occurred. In early 2002, China Unicom obtained permission from the central government to partly list its CDMA system on China's stock market.

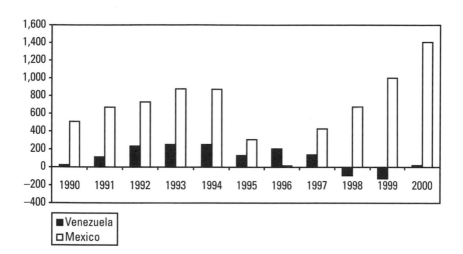

Figure 7.2 Growth of new subscribers for local fixed services after privatization. (*Source:* ITU International Telecommunication Indicators Database.)

This move may, perhaps, not be extravagantly interpreted as marking a critical step towards further radical and discontinuous policy change.

7.6.4 Legal Framework

To date there has been a complete absence of a Telecommunications Act in China. The lack of a codified law threatens to render regulation conflictual, complicated, inconsistent, and uncertain. The controversy over China Unicom's CCF investment scheme is illustrative of emergent problems. Through the CCF scheme, China Unicom has successfully invested in 23 projects with a total investment of $1.5 billion. In 1999, however, the MII unilaterally declared that the scheme is an unacceptable investment scheme, although it could not define it as illegal in the absence of a well-defined law regarding foreign direct investment in telecommunications. All China Unicom's partners were immediately required to withdraw their investments. This has seriously damaged the reputation of the Chinese telecommunications sector and produced uncertainty among foreign investors. This case, the exemplar of others, illustrates the urgent requirement for an effective and transparent legal framework.

Entry to the WTO provides China with a perfect opportunity to adjust its legal framework. By committing to the Regulatory Reference Paper, China would need to transform its traditional pattern of sporadic ad hoc intervention through the adoption of a transparent legal and regulatory framework. Although this might be viewed as deeply threatening within the context of the Chinese telecommunications framework, successful experience from early-mover countries certainly provides a policy lesson for China in setting up an efficient legal framework in a relatively short period of time; a policy lesson, moreover, founded on the declaration that a telecommunications legislative framework is a *sine qua non* of effective regulation. The MII's regulations on network interconnection published in 1999 [30] borrowed many principles from legislation governing the activities of the FCC and Oftel. If these regulations had been published immediately after China Unicom was established, there would not have been so many controversies over network interconnection and China Unicom would have been launched with much greater velocity.

In September 2000, a critical step was taken by the MII by publishing "Telecommunications Regulations of the People's Republic of China" (see Appendix C). Although this regulation remains at best an administrative document lacking legal status, it has nevertheless provided the industry with relatively transparent guidelines for operating the telecommunications business in China. The creation and sustenance of a regulatory framework embracing the principles of predictability, transparency, and fairness—all essential to the creation of a level playing field on which contestable market behavior may take place—suggests that, in the wake of WTO accession, early priority should be given by the People's Congress to the refinement of this regulation and its passage into a formal telecommunications law.

7.7 Conclusion

Although China has experienced a much lower telephone penetration rate than developed economies, there have been strong and seemingly irresistible political, economic, and financial pressures for China to join the WTO. Arguably, the telecommunications sector had already been transformed into a buyers' market with liberalization high on the change agenda before entry. China's application to accede to the WTO signaled the country's intent to extend this important policy experiment by further adoption of the norms and values of contestable markets through cautious embrace of aspects of a globalized agenda. The above account indicates that the last two decades of the twentieth century constituted a continuing long march of telecommunications reform. WTO accession can be seen as the logical capstone to a policy of lengthy internal reform and change in the telecommunications sector.

In spite of clear hesitations, which have been outlined in this chapter, the Chinese government manifestly comprehended the ineluctable pressures for WTO membership and eventually adopted a proactive stance in facilitating China's entry into the WTO. Historical experience, however, has imprinted on government consciousness the necessity of retaining control over telecommunications operations. The economic centrality of this industry has made the government extremely coy about opening it up to foreign incursion. The retention of a cap on foreign direct investment is one of the safeguards that it has sought to retain. Pragmatically, the government has attempted to adopt a middle-way approach by taking advantage of foreign investment on the one hand and securing its control over telecommunications operations on the other. A commitment to "dancing with wolves" is arguably the most effective strategy currently available to the Chinese government and operators. Such choreography involves closer relationships with partners held at arm's length!

The fact that even China—traditionally viewed as exceptionalist and autarkic—can be the beneficiary of foreign direct investment through WTO membership has wide implications. This membership surely implies that the WTO is not merely a club for the rich. The policy lesson here is that, for developing economies, the time is now ripe to commit to the WTO's Basic Telecommunications Agreement. Although similar concerns have been raised in developing economies to those in evidence in China—the issue of vulnerability to foreign incursion, for example—the experience of early-mover countries has shown, to an important degree, that these concerns have been greatly exaggerated. In fact, developing economies could enjoy even greater privileges than those experienced in fully developed economies through utilizing foreign investment. Imperfect competition in developing economies may even actually enhance the advantages of foreign direct investment [31].

Nonetheless, there is a clear message to be divined from China's experience. To fully exploit the benefits of joining the WTO club, China needs to drastically restructure its regulatory, legal, and organizational frameworks. Otherwise, as argued above, a chaotic playing field, an inconsistent policy stance, and a distorted market will seriously frustrate foreign investors. None of this would auger well for a Chinese public armed with rising expectations about telecommunications services. It would be a huge disappointment if the only beneficiaries of the foreign investment flowing into the Chinese system were the investor community. Chinese ambitions to turn the country into a major player in the international telecommunications arena depend upon a robust internal market and the diffusion of the benefits of telecommunications innovation to the entire population. Given the current skewed distribution of advanced telecommunications services to urban areas and the resultant digital divide, which remains to be addressed in this largely rural society, much remains to be done to secure levels of provision and universal service commensurate with China's development ambitions. The WTO provides the Chinese telecommunications system with an opportunity, but no guarantee, of future growth and dynamism. The continuing success of Chinese telecommunications will undoubtedly reside in the deftness of approach of the Chinese government in finessing the post-WTO policy environment. As the history of late-twentieth-century telecommunications has shown with its inherent contradictions and dilemmas, Chinese policy makers are well able to hold competing ideas in mind at the same time while still retaining the ability to function!

References

[1] WTO, "Data on Telecommunications Markets Covered by the WTO Negotiations on Basic Telecommunications," *WTO News Releases*, February 17, 1997.

[2] Oliver, C. M., "WTO Agreement on Basic Telecommunications Services and FCC Implementation," *Communications Lawyer*, Winter, 1998.

[3] Tarjanne, P., "Preparing for the Next Revolution in Telecommunications: Implementing the WTO Agreement," *Telecommunications Policy*, Vol. 23, No. 9, 1999, pp. 625–644.

[4] Coughlin, C., "Foreign-Owned Companies in the United States: Malign or Benign?" *Federal Reserve Bank of St. Louis Bulletin*, No. 74, 1992.

[5] Blomstrom, M., "Host Country Benefits of Foreign Investment," *National Bureau of Economy Research Working Paper*, No. 3615, 1991.

[6] Sidak, J. G., *Foreign Investment in American Telecommunications*, Chicago, IL: The University of Chicago Press, 1997.

[7] Ondrich, J., and M. Wasylenko, *Foreign Direct Investment in the United States: Issues, Magnitudes, and Location Choice of New Manufacturing Plants*, Kalamazoo, MI: W. E. Upjohn Institute for Employment Research, 1993.

[8] Graham, E. M., and P. R. Krugman, *Foreign Direct Investment in the United States*, 3rd ed., Washington, DC: Institute for International Economics, 1995.

[9] WTO, Annual Report of the Director-General, Geneva, Switzerland, 2001.

[10] "Zhu Rong-ji and China's Entering into the WTO," *Singtao Daily*, November 18, 1999.

[11] Wang, X. D., *Informationization: Choice of China in the 21st Century*, Beijing, China: China Social Science Publication House, 1998.

[12] Commentator, "Regulating the Internet Is an Anti-trend Behavior Against Liberalization," *Hong Kong Economic J.*, January 27, 2000.

[13] Zhang, H. L., "What Is WTO?" *China Economic Times*, November 16, 1999.

[14] Kan, K. L., "Where Is the Chinese Telecommunications Industry Going?" *Posts and Telecommunications Economic Management*, No. 10, 1999.

[15] Kelly, T., "Global Trends in Telecommunications Development," *Presentation at Hong Kong University of Science and Technology*, December 7, 2000.

[16] MPT, Provisional Arrangement for the Approval and Regulation of the Decentralized Telecommunications Services, 1993.

[17] He, X., "Study of P&T Financing Policies in China," *Research Report of the Ministry of Posts and Telecommunications*, 1996.

[18] Xu, Y., and D. C. Pitt, "One Country, Two Systems—Contrasting Approaches to Telecommunications Deregulation in Hong Kong and China," *Telecommunications Policy*, Vol. 23, No. 3/4, 1999a, pp. 245–260.

[19] Zhu, G. F., "Telecommunications Technology in China: Development and Prospects," *China Communications*, No. 27, 2000, pp. 14–18.

[20] Ginn, S. "Restructuring the Wireless Industry and the Information Skyway," *J. Economics and Management Strategy*, No. 4, 1995, pp. 139–145.

[21] Zhu, H., "The Impact of the Industrialization Policy on Telecommunications, the Deregulation Process and Foreign Financing: China's Case," *Communications & Strategies*, No. 41, 2001, pp. 37–61.

[22] Xu, Y., "China's Accession to the WTO and Its Implications for Foreign Direct Investment in Chinese Telecommunications," *Communications & Strategies*, No. 45, March 2002.

[23] Mueller, M., "China: Still the Enigmatic Giant," *Telecommunications Policy*, Vol. 18, No. 3, 1994, pp. 171–173.

[24] MII, "1999 Report of Communications Development," http://www.mii.gov.cn/news/dxggjh.htm, April 28, 2000.

[25] Liu, C., "Director of the Department of Policy and Regulations Comment on the Reform of Telecommunications in China," http://www.mii.gov.cn/news/dxggjh.htm, January 1, 1999.

[26] MII, "Banning the Irregular Price Competition in Cellular Market," http://www.mii.gov.cn/news/dxggjh.htm, November 3, 1999.

[27] Shi, W., "The Chinese Telecommunications Industry Needs To Be Significantly Restructured," *China Business Daily*, November 23, 1999.

[28] "Hong Kong Government Concerned About the Merger," *Singtao Daily*, January 26, 2000.

[29] "Telecom Talks Spurs Market to Record High," *South China Morning Post*, February 12, 2000.

[30] MII, Provisional Arrangement for Telecommunications Network Interconnection, Sepember 7, 1999.

[31] Hymer, S. H., *The International Operations of National Firms: A Study of Direct Foreign Investment*, Boston, MA: MIT Press, 1976.

8

Chinese Telecommunications Transformation: Cautionary Tales of a Paradigm Shift?

At the beginning of the twenty-first century, we are repeatedly assailed with prophetic announcements that the world is experiencing a dramatic process of discontinuous change. We are, in the views of many social critics, approaching a "climacteric"—summed up in the opinion of commentators, such as Alvin Toffler [1], that we are in the middle of a process connoted by a move from industrial to postindustrial society. The institutions that have been at the center of the industrialization process are now increasingly seen as ossified and redundant in a world of rapid change and social, political, and economic disjunction. The events of September 11, 2001 have served for some as a bellwether of a future world of unpredictability, change, and social stress. In the words of one American social scientist, "everything nailed down is coming loose" [2]. Information technology and telecommunications have both played key transformational roles in contemporary society. The key process of globalization, acknowledged by both its supporters and detractors as exercising increasing political and economic influence over nation states, owes its heightened role to a considerable degree to the transformative influence of communications technologies that have led to increasing interdependence between countries and erosion of the principles of national sovereignty. In a contemporary rehearsal of arguments, which surfaced as long ago as the 1950s, some observers claim that we are being "deafened by the din of collapsing ideologies" [3]. On this argument, technology is bringing about value change, which is itself productive of the erosion of ideological differences.

China has long fascinated observers. Its vast size and relative isolation from foreign influence have inspired both awe and suspicion. Napoleon's celebrated dictum that China was a sleeping giant continues to exercise influence over the Western mind, frequently leading to the conclusion that the country remains steeped in mystery. Chinese exceptionalism continues to hint that the country defies conventional categorization—its leadership alternately practicing authoritarianism while engaged in the deconstruction of central elements of its communist past. The telecommunications sector in China provides a crucible within which the blending of several (often competing and contradictory) models of political decision making coexist.

8.1 Policy Orthodoxy—Policy Heterodoxy

The past two decades have played host to discontinuous change in the structure and working practices of the telecommunications sector. Several impelling forces have been present throughout this period and have played a major role as the instigators of change. First, technological change has outpaced institutional development and put increasing strain on the dominant assumptions upon which orthodox telecommunications provisioning has been based. In particular, the bureaucratic structures and working practices typical of government organizations (and the PTT in particular) have been shown to be sluggish and unprepared for the explosion in demand for new and more efficient telecommunications services—a process that has clearly accelerated since the beginning of the 1980s. Second, ideological change has underpinned arguments in favor of a move away from statist provisioning of telecommunications to policies clearly favoring market allocation. This development was most clearly exemplified in the rise of Thatcherism in the United Kingdom with its clear message that "that government is best which governs least" in the economic sphere and elsewhere. Third has been the growth of consumerism—the rejection of the view that the recipient of government services was a mere subscriber whose views counted for little (in a world of monopoly provision) and its replacement with a model of the consumer as customer armed with better information and likely to demand higher qualities of service provision. Coupled with consumerism came the development of the idea of contestability—those markets would deliver greater allocative efficiency to the extent to which they could be opened up to competition. This message was driven home by policy influentials, such as the strategic decision makers in large businesses who increasingly complained of inefficiencies in the manner in which telecommunications services were delivered to their organizations. Finally, telecommunications came to be widely viewed, not simply as a boring utility function (plain old telephone service) but as both a key value chain component of modern business and a salient feature of private life. These changes appear to have been brought about by the convergence of telecommunications and computing technologies and the development of the so-called infocommunication industry.

Arguably, these forces together constituted increasing environmental uncertainty for telecommunications policy makers and were key to an understanding of the transformation of the dominant orthodoxy of telecommunications provisioning from that of monopoly/monopsony to an allocative strategy based on market ideology and greater competition. The shift in public values entailed by this move has been described as a move from administered to market governance [4].

Importantly, and as acknowledged in Chinese folklore, no idea is more important than one "whose time has arrived." Deregulation of telecommunications in Britain connoted an important policy experiment—one that was destined to command the attention of politicians elsewhere. Liberalization (the opening of the telecommunications market to competition) and privatization (the transference of state assets into private hands and the subjection of bodies so transferred to the pressures of market forces) acted as policy exemplars leading to policy transfer of these ideas into a wide variety of national policy domains. From Mexico to Manila, politicians studied this experiment in the deconstruction of familiar bureaucratic government structures in the effort to revivify the fortunes of their sometimes waning PTT and spread the virtues of ownership more widely.

Simply stated, change in the fundamental assumptions and working practices within the telecommunications sector in the last two decades of the twentieth century may be characterized as a paradigm shift—a fundamental (and potentially irreversible) rejection of a set of long-held assumptions in favor of the adoption of a set of radically different criteria. Thomas Kuhn, the architect of the term *paradigm shift*, characterizes advances in human knowledge as proceeding over long periods of time in a rather conservative manner until the system of ideas and values is rendered powerless to accommodate changes of a revolutionary nature [5]. Then, sufficient change pressures will force the abandonment of old ways of operating in favor of radical alternatives. Kuhn was quick to point out that such change might (at least for a time) be rejected by powerful social forces on the grounds that it is uncomfortable or too costly in emotional, intellectual, or financial terms. Eventually, however, pressure will mount to the point at which change becomes irresistible. In Kuhnian terms, the changes impacting on the fundament of the telecommunications provisioning regimes of the 1980s were sufficiently important and irresistible as to constitute a revolutionary period of transformation.

In previous chapters the point has been made and reinforced that China has not been impervious to changes clearly impacting not only the British system, but elsewhere. Alongside the British policy experiment was an important series of judicial and political processes in the United States that led to the break up, or divestiture, of the venerable AT&T, whose erstwhile president, Theodore Vail, had once boasted that the company provided "end-to-end service" and was an institution central to the American way of life. Such views did not prevail in the 1980s and 1990s, which saw the monolithic company forced to compete with its former Bell company constituents. These twin developments were flattered by policy emulation elsewhere. Deregulation became an industry in its own right in countries throughout the world. It appeared to be—to borrow an expression from another context—a situation of permanent revolution in the domain assumptions of telecommunications public policy orthodoxy.

China's telecommunications policy path has been briefly traced above. The early period of Chinese telecommunications (pre-1949) was one of fragmentation, stagnation, and foreign intervention resulting from invasion and civil war. The latter part of the nineteenth century witnessed the de facto takeover by foreign companies (such as the Danish Great Northern Company). The enduring policy lesson of this period (which still exercises influence over thinking in the telecommunications sector) was that foreign companies were essentially exploiting the system for their own commercial ends while paying scant regard to the public good. After the Communist takeover in 1949, China embarked on a process that enshrined the values of administered governance. The strategic military importance of the telecommunications system was recognized, as was its place as an integral part of a socialist centrally planned economy. Added to the normal problems of overcentralization in this period were the perils of turmoil resulting from the so-called Cultural Revolution. This severely disruptive sociopolitical process led to administrative suboptimization—that is, a telephone famine, which by 1980 had resulted in a penetration rate (0.43%) ranking amongst the lowest of 140 leading countries.

Remarkably, China recovered from this situation and in 1978 embarked on a process of truly extraordinary telecommunications reform. By the end of the 1990s, the country's telephone penetration rate had risen to over 20.1% with the total capacity of the PSTN amounting to no less than 179 million lines.

8.2 Propellants of Change

The same three factors that have propelled policy change in telecommunications regimes in first-mover countries (e.g., United Kingdom, United States) have arguably been present in China [6]. Radical technological change and, in particular, the convergence of telecommunications and computing has occurred in the Chinese telecommunications sector. In fact (and ironically) China has benefited from its relative technological backwardness—the contemporary Chinese telecommunications system had to be built upon virgin soil. This has, in fact, allowed China to gain the advantages of late-mover status and leapfrog to a position of technological sophistication. With the accelerator of a deregulated equipment policy, the contemporary Chinese public telecommunications network is in 2002 one of the most technologically advanced in the world. By 2001, 100% of the telecommunications system for both fixed and wireless had been digitized.

Second, telecommunications in China, as elsewhere, has become a key corporate resource as Chinese companies, like their Western counterparts, have been forced to recognize the heightened importance of communications as a central component in the value added chain of their business. China's efforts to play a full part in the world trading system have accentuated the importance of telecommunications in a commercial environment. Coupled with this is the unleashed demand for greater telecommunications connectivity on the individual level. This combination of commercial and individual pressures for change has played a major role in moving telecommunications from a supply-led to demand-driven provisioning regime.

Third, as elsewhere, China has played host to the ascendancy of a market-driven political economy. The Chinese government has been determined to transform its highly centralized and planned economy into a so-called socialist market economic system—and this in response to policy lessons from abroad. Driven by pragmatism, such "heresy"—itself evidence of fundamental change—is now routinized and increasingly appears to be seen as politically unexceptional. Deng Xiao Ping's mantle of change has fallen onto the shoulders of his successors.

The telecommunications sector has been a beneficiary of this broader process of change, with the Chinese government clearly committed to the acceleration of telecommunications development. In 1978, the *annus mirabilis* of Chinese telecommunications witnessed a sea change in the attitude of key participants in the telecommunications policy process: the commitment of the MPT, for example, to rapid infrastructural development, the inception of an open-door policy for inward investment, and the relinquishment of the government view that telecommunications should be exploited for government use rather than seen as a commodity, are all examples of the ascendancy of a new telecommunications order in China with its roots at least partially in policy developments elsewhere. As noted earlier, the arrival of China Unicom may be justifiably portrayed as an important component of a paradigm shift in the evolution of the Chinese telecommunications system. It signaled a clear commitment on the part of the country's policy makers to break with traditional bureaucratic (monopoly) patterns of administration and service delivery and engender economic liberalization and pluralistic (semicompetitive) provision presaging further attempts at structural deregulation elsewhere in the Chinese economy.

It is fascinating to note that cracks in the monolith of the centralized administered governance structure of Chinese policy making in the telecommunications sector were opened, perhaps surprisingly, by its erstwhile guardians. Key government departments, such as the MEI, came to see themselves sidelined by the MPT. Convinced that it was being excluded from MPT equipment contracts, the MEI pressed for the development of alternative power centers over which it might exercise greater influence. The support of the MEI for the liberalization of the sector was paralleled by that of the PLA. Seeking stable sources of income from which to finance its increasing overheads (pay and pensions, for example), the PLA shared the MEI's conviction that the dethronement of the MPT was an essential element in its efforts to become an activist player in this potentially colossal marketplace.

The activities of the MEI and the PLA as change agents provide a clear indication that key players recognized that telecommunications had enormous growth potential. Arguing that the MPT was failing to meet growing demand for telecommunications services, other ministries (such as the MOR), operating their own private networks, insisted that a policy of liberalization would free them to take up slack demand unsatisfied by the dominant provider (MPT). This was a potent argument, indicating that the government would do well to encourage these ministries to redirect their private networks to the provision of public telecommunications services.

Such arguments gained added weight in a system that, to this day, is still far from maturity. At the beginning of the twenty-first century, the

Chinese telephone penetration rate is only a twentieth of that of many developed countries—a position skewed by continuing underdevelopment in the rural area. Continuing high and unsatisfied demand has had an important twin impact on the sector. First, it puts MPT under considerable pressure, leading, in turn, to arguments in favor of breaking up its monopoly. Second, the very size of this underdeveloped market still offers vast potential for new entrants. Such characteristics of an immature market offer equally compelling reasons for deregulation as forces operating in more mature counterparts.

Some years ago an American scholar with an interest in China declared the conditions for revolution (a social paradigm shift) to be the generation of multiple dysfunctions in a society coupled with the presence of an accelerator in the form of a change agent [7]. It is perhaps not too extravagant to claim that such factors have been present in the Chinese telecommunications sector in this most intensive period of change. Chinese exposure to the world marketplace through the WTO and clear indications of pressure emanating from growing domestic demand threatened to destabilize the cozy monopolistic world prior to the 1990s. The emergence of China Unicom onto center stage during that decade was evidence of potentially seismic shifts taking place in the telecommunications provisioning regime. Such fundamental change did not occur spontaneously. On the contrary, echoing the Chinese dictum that "a time of crisis is a time of opportunity," key "private" network operators, such as the MOR and the MEP, saw their chance to correct undercapacity in the telecommunications sector by urging the State Council to allow them to compete with the MPT. To a considerable degree they acted as the midwives, or accelerators, of a telecommunications revolution.

To summarize, inspection of the developments in the Chinese telecommunications sector during the closing decades of the twentieth century reveals the presence of all the commonly recognized conditions that have led to telecommunications deregulation in first-mover countries (including technological development, a growing consumerist presence, and political impulsion). Additionally, the continuing presence of an immature (and huge) market coupled with the presence of strong private networks has provided a significant impeller of liberalization and deregulation.

The extraordinary pent-up demand for telecommunications stimulated by a rapidly developing economy, the existence of large-scale private networks, the resentment of the public towards the MPT, and the growing financial strength of China Unicom all have worked to help China overcome the characteristic disadvantages experienced by second-mover nations. China's supposed backwardness has, ironically, worked to its advantage.

Additionally, adroit development of the "one country, two systems" policy, under which elements of state socialism remain intact but coexist with social market developments, has effectively promoted China's coastal rim as a policy laboratory. This has showcased China's commercial potential while entailing the necessity of developing state-of-the-art communications.

8.3   Conferred Advantages?

The above account has attempted to document the obvious benefits to China of a policy of market liberalization in telecommunications. Such a development, which marks a major process of political transformation, has produced clear benefits.

8.3.1   Services and Manufacturing

Manifestly, introducing contestability to the telecommunications system has triggered new developments in service provision, many of which must be considered as explosive. The case of mobile illustrates the point. By the end of 2001 the mobile marketplace comprised the staggering figure of 145 million subscribers. The enormous growth in SMS services reveals the huge potential of the wireless data communication market in China—many new players have already moved in to provision demand or are poised to do so. China is now well on the way to deploying a state-of-the-art 3G mobile communications platform, possibly with China's 3G standard (TD-SCDMA) that has been approved by the ITU. Value added information services are now beginning to play an important role, such as in mobile banking, stock trading, and e-mail. As indicated above, mobile data roaming capacity has been developed and customers are increasingly using mobile as a supplement to the PC and sending and receiving e-mail and text messages.

Coupled with the increasing variety in provision of services has come more aggressive manufacturing activity with Chinese industry now contributing to important technological developments, such as SPC switching. The adoption of the 3G standard is already stimulating manufacturers in China to look well beyond 2G mobile communications.

The concept of managed liberalization probably best describes the role of the Chinese government in this period. From an insular protectionist stance, the government moved in the 1980s to what has been described as "import, digestion, absorption and creation"—adapting the skills and capacity of foreign manufacturers and using that as a basis for the development of indigenous manufacturing industry. Now a joint venture policy is well established. Vitally, increasing domestic manufacturing activity has intensified competition in the equipment market and forced foreign manufacturers to reduce their prices.

It has been argued above that the attenuated process of accession to the WTO seems destined to force an additional round of reform in the telecommunications sector measured in yet more acceleration of the pace of transition in the industry and an intensification of now well-established market trends.

8.3.2 Organizational Capacity

To echo the words of the organizational theorist, telecommunications in the Chinese context have mirrored change pressures in other telecommunications domains. A secular trend in the global telecommunications industry has been rapid change—it has moved from a placid, reactive environment to one of unpredictability and turbulence. These environmental factors have impacted on telecommunications organizations worldwide.

Such environmental forces have typically brought about a review of the structure and working practices of pivotal telecommunications organizations. A useful characterization of such impacts can be viewed through the useful distinction between mechanistic and organic organizational models [8].

The former denotes a characteristic bureaucratic organization, with a well-developed hierarchy, heavy reliance on rules, and lengthy lines of upward/downward communication. The latter describes a much flatter organization, informally based and adaptable. The mechanistic organization appears most useful for routine administration in conditions of environmental stability. Research clearly indicates that the organic form is likely to be utilized where an organization needs to adopt a rapid response capability to changing, turbulent, and unpredictable circumstances. The history of telecommunications during the 1980s and beyond is one in which many of the PTTs throughout the world attempted to change from a mechanistic to organic structure more in keeping with changing times [9].

The Chinese organizational experience suggests that the organizational infrastructure of its telecommunications sector has not been immune from changing environmental pressures, and it, too, has made efforts to bring its (largely mechanistic organizations) into line with changing reality.

Clear indications of the impact of the change process on Chinese telecommunications organizations suggest that they, like their overseas counterparts, have attempted, and continue to attempt, to put (more organic) structures in place to better enable the organizations to respond quickly and more effectively to the change pressures impacting on them.

The period from the late 1940s until the 1970s was one characterized by highly centralized structures and a monopoly, which discouraged risk taking and the assumption of responsibility at low levels in the organization. Such an arrangement was inimical to a transformation from a mistake avoiding strategy to one of greater risk taking.

Low down on the list of state priorities until the 1980s, telecommunications was seen as an instrument of governmental administration and national defense. Continuing problems of overcentralization (with attendant bureaucratic control) were exacerbated by the funding regime within which telecommunications was administered. In a situation reminiscent of conditions facing the British Post Office during the 1950s, revenues reverted to the central government rather than being hypothecated for the development of the telecommunications system. This situation was partially remedied in 1969 with the "hiving off" from posts of the telecommunications function in Britain and its relocation in a separate government corporation.

Finally, in 1982 strenuous efforts were made to change the funding regime. Henceforth, government gave priority to the MPT in funding, foreign currency allocation, and foreign government loans. Such marked the end of the period of capital famine. As has been argued above, this process was further enhanced from the mid 1990s with radical changes in the funding regime.

With the Chinese economy migrating to market economy status, the fiscal and monetary environment within which telecommunications was operating experienced radical change. For example, the government allowed faster depreciation of fixed assets to encourage the development of network renovation and levied a lower profit tax on the industry.

In regard to foreign investment, restrictions were effectively circumvented through Hong Kong. A company like China Mobile was able to use Hong Kong to attract foreign investment and reinvest it in China proper as such capital was not considered to be foreign direct investment. Similar results were obtained through use of the stratagem of the CCF investment scheme.

Finally, and importantly, a process of incentivization has been in operation in the last two decades. Consistent with changes in organizational structure have come changes in human resource management. Efforts were eventually made through the implementation of the "difference" management system referred to earlier to enable firms in the sector to keep a proportion of their surplus in an attempt to motivate individual enterprise managers to consider efficiency improvements as a major part of their corporate strategy. The introduction of a Contractual Responsibility System and the so-called Director Responsibility System were instigated with the twin objectives of decentralizing power and responsibility to individual enterprises (within the overall boundaries of a government planned economy structure) and motivating key decision makers (directors) through individualized rewards, such as promotion (a clear move away from the iron-rice-bowl model of remuneration of earlier years). A process of salary reform put into operation since 1985 has resulted in an emphasis on the notion of responsibility and autonomy. Of no lesser significance has been the awakening of the idea that the recipients of telecommunications policy should now be regarded as activist customers with a legitimate voice, which can be raised to demand high qualities of service delivery and complain when such are lacking.

The merits of such a shift in mindset appear to be clearly measurable efficiency improvements and increased labor productivity. Confidence in such measurability has increased with the adoption of more robust economic accounting systems allowing for improvements in the quantification of results. This has resulted in better indications of true profit and loss and the operationalization of schemes to measure profitability—all consistent with a more commercial approach to the telecommunications business.

In brief, the past two decades (and longer) have played host to major changes in the working practices of organizations within the Chinese telecommunications sector. The encouragement of competition and market entry has resulted from a deliberate attempt to stimulate change, innovation, and risk taking. The telecommunications sector has moved from a situation of centralization and stability to a more polyarchal system. Remarkable enough within the context of the settled frames of reference of first-mover countries, this can be seen as quite exceptional within the Chinese context. The fundament of the centrist, state-run bureaucratic system has been effectively shaken root and branch—the "assumptive world" of Chinese telecommunications has, to an important degree, been the subject of a paradigm shift. Patterns of working, organizational structures, financial arrangements, and the redistribution of authority and power through a decentralized decision-making system are all parts of the legacy of reform.

The long march of Chinese telecommunications reform seems set to run its course for a considerable time to come. China's accession to the WTO has already concentrated the minds of decision makers within the telecommunications domain and appears set to do so in the future. There has long been recognition among the political and managerial elite that China could reap heady rewards through the gateway to the world provided by WTO membership. On the other hand, sensible preparations have had to be made to meet the potential onslaught from foreign competitors in key Chinese markets—one such is telecommunications. The reconfiguration of the Chinese telecommunications marketplace and the repositioning of key organizations within the sector is testimony to the foresight of Chinese decision makers in recognizing the need for fundamental reforms in operation and working practice.

In summary, it is clear that telecommunications deregulation has already brought about very positive changes within the Chinese telecommunications sector. It is not too extravagant to claim that a revolution has occurred in the Chinese telecommunications industry. The benefits achieved so far will undoubtedly promote further developments in telecommunications deregulation and further strengthen the determination of the Chinese government to liberalize its economy. It is also evident that, for a socialist country like China hosting a distinctive political and economic system, telecommunications deregulation has proved a necessary and feasible public policy paradigm.

8.4   A Paradigm Shift with Chinese Characteristics?

The present account has been at pains to indicate that the closing decades of the twentieth century and the opening years of the twenty-first have provided a stage upon which an unfolding drama of change has been acted out. There are clear indications that the Chinese telecommunications system has been buffeted by change pressures and has responded to them by discarding many of the myths and shibboleths taken for granted by observers of the Chinese scene. Examination of Chinese telecommunications provides an insight into the workings of the Chinese politico-economic system as it has attempted to come to terms with the forces of change engendered both within and outside the country. *Cui bono?* For clients or customers of the system, clear benefits would appear to have accrued: a wider range of services, reduced prices, and more efficient working. China can rightfully claim that the cumulative benefits of change are clear and unarguable. Its citizens have reaped the rewards of change just as have those of other countries like the United Kingdom and the United States.

Policy Paths and Policy Transfer

Several years ago the idea of technological convergence became very fashionable; this is the idea that, regardless of political and economic differences, countries were converging politically through a process of technological determinism. Technological advances were blurring ideological and cultural boundaries and bringing about a process of political and economic uniformity.

It is tempting to suggest that technological change in the telecommunications sector is having a convergent impact and bringing about such uniformity in the working practices of the industry. The Chinese experience is certainly testimony to the significance of technological drivers of change. It also indicates that the Chinese government has not been slow in learning the policy lessons of developments elsewhere and adopting policy transfer of ideas, such as liberalization within the Chinese context. Uncritical acceptance of this view, however, would obscure very real differences in the operationalization of concepts, such as liberalization and privatization within China.

The concept of policy path (a term used by contemporary political scientists) is useful in alerting us to the possibility that the same set of forces may produce variance in outcomes as it is channeled through the politico-cultural fabric of different countries. As policies proceed through the formulation and implementation phase, they will be configured and modified subject to the political and cultural nuances of the host political system [10].

The most obvious outcome of this observation is that, while China may be said to have learned important lessons from afar, those lessons have not been put slavishly into operation in the attempt to completely mimic them. The most obvious example is that of privatization. The bold (and arguably successful) policy of privatization of BT has done much to proselytize the idea that privatization and deregulation go hand in hand. Privatization is no longer to be seen as "a fashionable concept or an empty slogan, but rather as a key strategic tool for telecommunications business" [11].

While privatization proved to be an idea whose time had come, it still has little political or cultural resonance in China, which, in spite of bold policy experimentation, remains a socialist country and one in which state-owned enterprises still occupy dominant positions in the national economy, particularly in key infrastructural sectors. Privatization is simply absent from the contemporary Chinese political agenda, symbolizing as it does a commitment to capitalist modes of enterprise management.

This account raises the key issue of whether, in the absence of full-blown privatization justified on the argument that it is a prime vehicle of risk taking entrepreneurship, the citizens of China will benefit from a controlled liberalization stopping short of transferring state assets into the hands of private sector entrepreneurs.

In fact, this book demonstrates that, for countries where privatization remains as an unavailable policy option, competition can still be introduced successfully on the government's policy agenda and that such competition can bestow clear benefits on the wider community. There are three supports for this contention.

First, contemporary Chinese experience teaches that state-owned operators, to an important extent, can achieve similar benefits to those generally ascribed to privatization through important internal reform. Since 1978, China has embarked on a process of reform of its economic system, its business accounting system, and the adoption of incentivization schemes and large-scale schemes of enterprise decentralization, which have provided the bedrock of an effectively working commercial system.

Second, experience in the telecommunications sector over the past decade has demonstrated that state-owned operators are in a position to respond to competition similar to privatized Western firms. Considerable evidence can be adduced that clearly shows that both the incumbent and the new entrants have reacted to competition swiftly and proactively, especially within a regulatory framework. They appear to have evolved from mechanistic to organic organizational types. All operators have fully utilized their strength to defend their respective positions and penetrate the market. Volume growth in markets, such as mobile telecommunications, suggests that state organizations have transformed themselves into formidable competitors in a rapidly changing market environment

Third, clear evidence exists that the public can benefit from competition between two state-owned operators. Following the inception of China Unicom in 1994, for example, the public has manifestly benefited from competition between the two rivals. Such benefits have accrued from elimination of installation fees and reduction in tariffs (most obviously in mobile and long-distance service) as well as obvious improvements in service delivery.

Such evidence suggests that state ownership is not necessarily an insuperable barrier to the introduction of competitive mechanisms. State-owned operators can be effective in a competitive market so long as proxies and measures can be devised that mimic aspects of privatization. In fact, the only constraint seems to come from the regulator's concern over the rapid and uncontrolled devaluation of government assets resulting from intense head-to-head competition. Such regulatory protectionism (manifested in such schemes as the construction of price floors) has clearly failed to impress market players. Predictably, they have busied themselves discovering the means to bypass such kindly proffered but intrusive regulation.

The Chinese telecommunications policy experiment reveals that deregulating telecommunications in the absence of privatization is, indeed, a feasible policy option and one, moreover, that does not cause undue disturbance to the norms and values of a socialist political system. Such a conclusion is of no small significance. The whole question of telecommunications reform and privatization must be contingently related to national culture and sociopolitical environments. This observation indicates that, in countries where fully developed privatization schemes are still not available as policy options, China proves to be a more apt policy exemplar than first-mover countries in the developed West. The central policy lesson from China is that telecommunications deregulation can still be used as an effective way of promoting the development of telecommunications. The competitive paradigm can be adapted to a wide variety of circumstances.

It is very important, however, to emphasize that competition without privatization is a feasible but not necessarily optimal policy. As was pointed out in Chapter 7, in order to achieve the full benefits of competition and to avoid potential interest conflicts in the post-WTO era, the privatization of telecommunications should be elevated as a high priority on the government's agenda when ideological barriers to its implementation are finally removed.

8.5 Telecommunications Deregulation: Limits to the Possible?

Writing more than a decade ago and in the wake of the telecommunications deregulation and privatization movement impacting across the globe, Trauth and Pitt [6] posed the question, are there limits to the successful implementation of the deregulatory paradigm? China provides the location for an interesting policy laboratory in which this question may be destructively tested. Unsurprisingly, examination of China's telecommunications paradigm shift reveals that its implementation has not been entirely untroubled.

One of the serious issues emanating from this process has been that of the dual status of the MPT as both an operator and a regulator. The close affinity between China Telecom and the MPT meant that it has proved impossible for the MPT to regulate the Chinese telecommunications market in an equal, fair, and equitable way. Whenever there has been a dispute, China Unicom has had to approach the State Council to have the problem resolved. This has attenuated the decision-making process and has led to the specter of regulatory dualism and confusion. Throughout the past few years the government has shied away from radical reform of the regulatory structure, instead opting for a gradualist approach to the introduction of competition and an independent regulatory authority. The position of the regulatory authority is further complicated by the absence of a statutory framework for telecommunications within the Chinese system. Failure to pass a national

Telecommunications Act has effectively led to the continuation of a disputes resolution process in which the State Council is asked to adjudicate between, for example, China Telecom and China Unicom on a case-by-case basis. This contrasts markedly with the situation in the United Kingdom, for example, where Oftel (an independent regulator) has operated under the authority of the 1984 Telecommunications Act and is armed with comprehensive means of enforcement, such as amending or revoking operators' licenses. In the United States, according to the 1996 Telecommunications Act and other pieces of legislation, disagreements between the FCC and telecommunications entities may be heard in a civil court interpreting matters at hand. Failure to draft a comprehensive Chinese telecommunications law has been attributed to "agency competition" between the various government departments concerned—a process that has been characterized as "contentious and possibly rancorous."

This legislative gap has been productive of clear policy deficiencies. It lays open the possibility of government by fiat, a system moreover subject to policy flip-flops dependent on the personality and preferences of particular members of the State Council. Lack of a legislative framework, therefore, raises the possibility of inconsistency, unpredictability, and capriciousness. Additionally, key policy issues may lay unattended or underemphasized. Two such issues are interconnection and universal service. An indeterminate regulatory system and lack of a clear telecommunications legal framework could render interconnection a key future bottleneck issue in the Chinese context. In contrast again to the situation in the United Kingdom and the United States, absence of a legislative universal service obligation (USO) requirement has had deleterious consequences for national telecommunications provision. There has been a skewed distribution of telecommunications services to major cities and the wealthier coastal strip. By the end of 2001, for example, while the telephone mainline penetration rate had reached 20.4% in urban areas, 93.2% villages were totally devoid of basic telecommunications service [12]. Arguably, the digital divide between the so-called information rich and information poor has widened as a result of competition. The absence of a legislative universal service obligation on operators has led them to concentrate on areas of higher wealth rather than the rural hinterland. In a familiar game of leaving the poor to catch up, the major benefits of competition have been conferred on a relatively rich constituency. As of 2002, the unintended distributional consequences of competition remain obdurately unattended.

Entry to the WTO will surely test the integrity of the Chinese policy experiment. It is unlikely that the Chinese government will suddenly open its markets to foreign incursion and encourage inward bound competition without let or hindrance. Indications of this were apparent in the speech delivered to the prestigious Pacific Telecommunications Council meeting in Hawaii in early 2002. Pointing to the vast potential immanent within the Chinese telecommunications and information technology marketplace, Wu Ji-chuan, Minister of the Information Industry, said that changes in the Chinese system consequent upon WTO entry would be both "deliberate and orderly." He insisted that the Chinese market "must be opened up and regulated in an orderly manner."

Such cautionary words signal concerns within the telecommunications policy community that China could suffer from asymmetry—opening up its potentially lucrative domestic market to foreign equipment and service providers while its indigenous manufacturers and telecommunications companies find it difficult to achieve overseas market penetration. Wu's pronouncements once again lend weight to the argument that China will in the future, as in the past, fail to encourage a great (policy) leap forward but will opt instead for the tried and trusted incremental pursuit of touching stones to cross the river. To what extent this incrementalist line can be held in the face of exponential competition remains for the present an open question. As this book reaches its conclusion, the announcement has just been made that British Telecom has signaled the consummation of its new relationship with a Chinese partner, 21 Vianet, which will open access to the Chinese value added network segment. On March 22, 2002, Shanghai Symphony Telecom, a joint venture between China Telecom (60% ownership), AT&T (25% ownership), and Shanghai Information Investment Inc. (15% ownership), formally launched its broadband service in Shanghai. Such are surely harbingers of a potential deluge of such deals in the wake of WTO. The resonance of the venerable Chinese phrase *a time of crisis is a time of opportunity* retains its contemporary significance. To turn the phrase around, opportunity may contain the seeds of future problems. Privatization may unleash the potential of Chinese companies into the international business arena. Equally, these debutante companies may be particularly vulnerable to the competitive threat for existing internationally experienced incumbents. Government policy makers may be forced to grapple with the intended and unintended consequences of a policy of permanent privatization of Chinese state assets.

No less than in China, experience in first-mover countries has shown that the path of telecommunications deregulation is long and difficult and strewn with contention and argument. In China, the existence of an indeterminate regulatory structure coupled with the lack of a robust legal framework for telecommunications and the consequences, foreseen and unforeseen, of WTO entry renders prediction difficult. Given China's historical provenance, however, its embrace of telecommunications deregulation, characterized as halting only by the most partisan critics, can be presented as a hugely important policy commitment. Telecommunications reform in China may be part of an unfinished revolution whose completion may be difficult to predict, but there seems little doubt that it reflects the wider dynamics of change in China—a society deeply immersed in the throes of transition.

References

[1] Toffler, A., *Future Shock*, London, Engand: Pan Books, 1970.

[2] Bennis, W., *Beyond Bureaucracy*, New York: McGraw Hill, 1973.

[3] Singleton, N., and D. Singleton, *Paradise Dreamed*, London, Engand: Bloomsbury, 1993.

[4] Huntley, J. A. K., and D. C. Pitt, "Transformation in Telecommunications Governance: From Administrative to Market Approaches," paper presented to QUT Key Centre, Brisbane, Australia, 1993.

[5] Kuhn, T., *The Structure of Scientific Revolutions*, Chicago, IL: Chicago University Press, 1962.

[6] Trauth, E. M., and D. C. Pitt, "Competition in the Telecommunications Industry: A New Global Paradigm and Its Limits," *J. Information Technology*, Vol. 7, No. 1, 1992, pp. 3–11.

[7] Johnson, C., *Revolution and the Social System*, Stanford, CA: Hoover Institute Press, 1966.

[8] Burns, T., and G. M. Stalker, *The Management of Innovation*, London, Engand: Tavistock, 1961.

[9] Pitt, D. C., and K. Morgan, "Bureaucracy, Deregulation and Technology: The 'Ramping of British Telecom and AT&T'," *Business in the Contemporary World*, Winter 1989.

[10] Xu, Y., and D. C. Pitt, "One Country, Two Systems—Contrasting Approaches to Telecommunications Deregulation in Hong Kong and China," *Telecommunications Policy*, Vol. 23, Nos. 3, 4, 1999, pp. 245–60.

[11] Kok, B., "Privatisation in Telecommunications—Empty Slogan or Strategic Tool?" *Telecommunications Policy*, Vol. 16, 1992, pp. 699–704.

[12] MII, "Statistics of Telecommunications at December 2001," http://www.mii.gov.cn, January 2002.

Appendix A: Individual Department Functions of the Ministry of Information Industry

General Office

- Dealing with the routine work of headquarters, coordinating the relations among various departments, and holding key meetings of the ministry;
- In charge of the information and news release, public relations, documentation, security, and other administrative issues.

Department of Policy and Regulation

- Stipulating the development strategy of the information industry;
- Drafting laws and regulations;
- Monitoring implementation of the regulations;
- Formulating policies on foreign investment;
- Dealing with issues concerning the Hong Kong Special Administrative Region, Taiwan and Macao.

Department of General Planning

- Formulating the development strategy for telecommunications, electronic product manufacturing, and the software industry;
- Crafting medium- and long-term plans;
- Coordinating the development of the telecommunications network, computer information network, broadcasting network, and all private networks;
- Promoting cohesive development between public and private networks and between services and manufacturing;
- Optimizing the allocation of resources and preventing duplicative construction;
- Managing state investment according to national budget priorities, supervising foreign technology transfer, foreign investment, and the establishment of joint ventures;
- Formulating technical standards for network construction and system design;
- Producing statistics and publishing information.

Department of Science and Technology

- Tracing the latest technology developments worldwide;
- Stipulating the R&D program and policy;
- Formulating technical standards for public telecommunications networks and broadcasting networks;
- Coordinating key R&D projects and facilitating the industrialization of these projects;
- Monitoring the quality of IT products;
- Managing the standardization, auditing, and data mining of information technology.

Department of Economic Reform and Economic Operation

- Drafting reform schemes for enterprises;
- Supervising the strategic restructuring of state owned enterprises;

- Coordinating major issues raised by enterprise reform;
- Auditing, monitoring, and analyzing economic performance;
- Forecasting the major economic indices of development;
- Macromanagement of the IT product market;
- Coordinating the import of IT products.

Bureau of Telecommunications Administration

- Formulating telecommunications development strategy and policy;
- Regulating telecommunications and information services in accordance with the legal framework;
- Preserving fair competition and ensuring universal service;
- Defending the interests of the state and the subscriber;
- Examining applications and issuing licenses for telecommunication and information services;
- Regulating service quality and tariffs;
- Defining and implementing the terms of network interconnection and accounting settlement schemes;
- Allocating network codes, Internet domain names and addresses;
- Supervision of the identification of standards for network facilities and terminal equipment;
- Supervising and coordinating the construction of private networks specialized for governmental uses;
- Supervising the State Network Management Center, international gateway, and Internet Security Center;
- Arranging and coordinating telecommunications service in case of emergencies;
- Formulating regulations on network and information security.

Department of Financial Adjustment and Clearing

- Implementing state governmental regulations concerning the management of state-owned property;
- Formulating regulations and accounting systems specifically applicable to the telecommunications sector;

- Stipulating accounting schemes on revenue reallocation among individual enterprises;
- Coordinating economic relationships among the post and telecommunications enterprises;
- Arranging both universal service subsidization and postal service subsidization;
- Stipulating tariffs for basic postal and telecommunications services.

Department of Electronic Information Product Administration

- Formulating long-term development strategy and policy in the manufacturing of IT products;
- Restructuring production output;
- Organizing major R&D projects in system equipment and microelectronics;
- Promoting the localization of components, materials, and equipment used in the state's key projects;
- Promoting the application of information technology.

Bureau of Military Electronic Industry

- Charged with management of the professional management of the military electronic industry.

Department of Informationization Promotion (State Informationization Office)

- Formulating strategies on promoting the informationization of society and economy;
- Supervising the progress of informationization in different industrial sectors and different regions;
- Providing support to enterprise for key informationization projects;
- Propelling the development of the software industry;
- Stipulating policies for the development and application of information resources;
- Promoting the popularization of informationization.

Radio Regulatory Department (State Radio Office)

- Preparing an effective spectrum allocation plan;
- Regulating radio stations and radio detection; in charge of the allocation and management of spectrum;
- Coordinating issues on frequency interference;
- Regulating radio in accordance to legislation;
- Coordinating the position of satellite orbit;
- Attending World Radio Conference on behalf of the government and dealing with relevant issue with relevant countries.

Department of Foreign Affairs

- Participating in information-industry-related international organizations on behalf of the government;
- Organizing and coordinating the signing and implementation of intergovernmental agreements;
- Handling intergovernmental issues on telecommunications and information;
- Responsible for international technical and economical cooperation and exchange;
- In charge of the approval and management of international cooperative projects and personnel exchange.

Department of Personnel

- Responsible for personnel management in accordance with its authority;
- Responsible for the training of human resources and intellectual exchange;
- Responsible for salary system and industrial relations.

Appendix B: Provisions on the Administration of Foreign-Invested Telecommunications Enterprises[1]

(State Council of People's Republic of China, December 11, 2001)

Article 1.   To meet the needs of opening the telecommunications industry to the outside world, and to promote the development of the telecommunications industry, following Provisions are formulated in accordance with the relevant laws and administrative regulations governing foreign investment and the Telecommunications Regulations of the People's Republic of China (the "Telecommunications Regulations").

Article 2.   Foreign-invested telecommunications enterprises refer to those that are jointly established by foreign and Chinese investors in the People's Republic of China in the form of Chinese-foreign equity joint ventures in accordance with law and specifically for the provision of telecommunication services.

---

1. This document is translated with reference to the original version of the State Council in Chinese and the translated version of Paul, Weiss, Rifkind, Wharton & Garrison Co. in English. It should be used for general informational purposes and should not be construed as, or used as a substitute for, a legal document.

Article 3.   In addition to these provisions, foreign-invested telecommunications enterprises shall comply with the provisions of the Telecommunications Regulations and other relevant laws and administrative regulations in the provision of telecommunications services.

Article 4.   Foreign-invested enterprises can provide basic telecommunications services and value-added telecommunication services. The Telecommunications Regulations should be required to make specific distinction between the categories of telecommunications services.

It is up to the supervisory department for the information industry under the State Council to define the geographic scope in which foreign-invested telecommunications enterprises may operate.

Article 5.   The following provisions should be complied with by a foreign-invested telecommunications enterprise in regard to the registered capital:

1. To provide a telecommunications service nationwide or in several provinces, autonomous regions or municipalities directly under the central government, the minimum amount of the registered capital of a foreign-invested telecommunications enterprise shall be RMB2 billion in the case of a basic telecommunications service or RMB10 million in the case of a value-added telecommunications service.

2. To provide a telecommunications service within one single province, autonomous region or municipality directly under the central government, the minimum amount of the registered capital of a foreign-invested telecommunications enterprise shall be RMB200 million In the case of a basic telecommunications service or RMB1 million in the case of a value-added telecommunications service.

Article 6.   The capital contribution ratio of the foreign investor in a foreign-invested telecommunications enterprise for the provision of a basic telecommunications service (other than radio paging service) shall not exceed 49% ultimately.

The capital contribution ratio of the foreign investor in a foreign invested telecommunications enterprise for the provision of a value-added telecommunications service (including radio paging which is in the category of basic service) shall not exceed 50% ultimately.

It is up to the supervisory department for the information industry under the State Council, in accordance with the relevant regulations, to determine the capital contribution ratio between the Chinese and the foreign investor of a foreign-invested telecommunications enterprise in different periods of time.

Article 7. To provide a telecommunications service, foreign invested telecommunications enterprises shall, in addition to satisfying the conditions set forth in Articles 4, 5 and 6 of these Provisions, also satisfy the conditions for providing basic telecommunications services or value-added telecommunications services set forth in the Telecommunications Regulations.

Article 8. The following conditions are applicable to the principal Chinese investor in a foreign-invested telecommunications enterprise providing a basic telecommunications service:

1. To be a company that is established in accordance with law;
2. To have appropriate funds and professional staffs for operations;
3. To qualify for the prudential and industry-specific criteria defined by the supervisory department for the information industry under the State Council.

The principal Chinese investor in a foreign-invested telecommunications enterprise referred to in the preceding paragraph refers to the investor whose capital contribution accounts for over 30% of the total amount of capital contributions of all Chinese investors, and its capital contribution makes the largest portion among all Chinese investors.

Article 9. The following conditions are applicable to the principal foreign investor of a foreign-invested telecommunications enterprise providing basic telecommunications services:

1. Have enterprise 'legal person' status;
2. Have obtained a license for providing basic telecommunications service in the country or territory where the licensee has registered;
3. Have appropriate funds and professional staffs for operation;
4. Have a sound record and experience in providing basic telecommunications services.

The principal foreign investor in a foreign-invested telecommunications enterprise referred to in the preceding paragraph as an investor whose capital contribution accounts for over 30% of the total amount of capital contributions of all foreign investors, and whose capital contribution constitutes the largest portion among all foreign investors.

Article 10. The principal foreign investor in a foreign-invested telecommunications enterprise providing a value-added telecommunications service shall have a good record and experience in providing value-added telecommunications service.

Article 11. To establish a foreign-invested telecommunications enterprise, either providing a value-added telecommunications service across provinces, autonomous regions or municipalities directly under the central government, or providing a basic telecommunications service, the principal Chinese investor shall submit an application to the supervisory department for the in- formation industry under the State Council and provide the following documents:

1. A project proposal;
2. A feasibility study report;
3. Qualification and relevant certification documents of the investors of the joint venture as set forth in Articles 8, 9 and 10 hereof; and
4. Any other qualification and certification documents for providing basic or value-added telecommunications services as set forth in the Telecommunications Regulations.

The supervisory department for the information industry under the State Council shall examine the above documents from the date of receipt of the application. If the application is for providing a basic telecommunications service, it shall complete the examination and make a decision on either approval or rejection within 180 days. If the application is for providing a value-added telecommunications service, it shall complete the examination and make a decision on either approval or rejection within 90 days. Where approval is granted, an Examination Memorandum For Providing Telecommunications Service With Foreign Investment will be issued; where application is rejected, the applicant shall be notified with justified reasons in writing.

Article 12. To submit an application in accordance with Article 11 for establishing a foreign-invested telecommunications enterprise, either providing a value-added telecommunications service across provinces, autonomous regions or municipalities directly under the central government, or providing a basic telecommunications service, its principal Chinese investor may, depending on the actual situation, submit documents other than the feasibility study report first, and then wait and submit the feasibility study report on its receipt of the notification of acceptance by the supervisory department for the information industry under the State Council. However, the interval between the date on which the notice of acceptance is issued and the date on which the feasibility study report is submitted shall not exceed one year, and this period of time is excluded from the stipulated timeframe for examination and approval.

Article 13. To establish a foreign-invested telecommunications enterprise providing a value-added telecommunications service within one province, autonomous region or municipality directly under the central government, its principal Chinese investor shall submit an application to the telecommunications administration authority of the province, autonomous region or municipality directly under the central government and provide the following documents:

1. A feasibility study report;
2. Qualification and relevant certification documents as set forth in Article 10 hereof; and
3. Any other qualification and certification documents regarding other conditions necessary for providing value-added telecommunications services as set forth in the Telecommunications Regulations.

The telecommunications administration authority of the province, autonomous region or municipality directly under the central government shall give comment on either approval or rejection within 60 days from the date of its receipt of the application. For approval, it shall forward the application to the supervisory department for the information industry under the State Council. For rejection, it shall notify the applicant with justified reasons in writing.

The supervisory department for the information industry under the State Council shall complete the examination within 30 days from the date of its receipt of the application documents with approval comment by the telecommunications administration authority of the province, autonomous region or municipality directly under the central government and make decisions on either approval or rejection. Where approval is granted, an Examination Memorandum For Providing Telecommunications Service With Foreign Investment will be issued; where application is rejected, the applicant shall be notified with justified reasons in writing.

Article 14. The main content of the project proposal for a foreign-invested telecommunications enterprise shall include: company names and basic background of the parties to the joint venture, the total amount of capital to be invested, registered capital, ratios of investment of each party, categories of services for which the application is made and during of the joint venture.

The main content of the feasibility study report of a foreign-invested telecommunications enterprise shall include the basic description of the enterprise to be established, categories of services, service projections and development plans, analysis of investment returns and anticipated time for services to be launched.

Article 15. If, in accordance with the relevant provisions of the State, the investment project to establish a foreign-invested telecommunications enterprise must be examined and approved by the department in charge of planning under the State Council or the department for general administration of the economy under the State Council, the supervisory department for the information industry under the State Council shall forward the application documents to the department in charge of planning under the State Council or the department for general administration of the economy under the State Council for examination and approval before issuing the Examination Memorandum For Providing Telecommunications Service With Foreign Investment. In the circumstance that application is forwarded to the department in charge of planning under the State Council or the department for general administration of the economy under the State Council for examination and approval, the timeframe for examination and approval in Articles 11 and 13 may be extended by 30 days.

Article 16. To establish a foreign-invested telecommunications enterprise, either providing a value-added telecommunications service across provinces, autonomous regions and municipalities directly under the central government, or providing a basic telecommunications service, the principal Chinese investor shall submit the contract and the articles of association for the proposed foreign invested telecommunications enterprise to the department in charge of foreign economic relations and trade under the State Council on the strength of the Examination Memorandum For Providing Telecommunications Service With Foreign Investment; in the case of a foreign-invested telecommunications enterprise providing a value-added telecommunication service within one province, autonomous region or municipality directly under the central government, the principal Chinese investor shall submit the contract and the articles of association for the proposed foreign-invested telecommunications enterprise to the department in charge of foreign economic relations and trade of the people's government of the province, autonomous region or municipality directly under the central government on the strength of the Examination Memorandum for Providing Telecommunications Service with Foreign Investment.

The department in charge of foreign economic relations and trade under the State Council and the departments in charge of foreign economic relations and trade of the people's governments of the provinces, autonomous regions and municipalities directly under the central government shall complete the examination within 90 days from the date of receipt of the contract and the articles of association for the proposed foreign-invested telecommunications enterprise and make decisions on approval or rejection. Where approval is granted, a Foreign-Invested Enterprise Approval Certificate will be issued; where application is rejected, the applicant shall be notified with justified reasons in writing.

Article 17. The principal Chinese investor in the foreign-invested telecommunications enterprise shall apply for an Operation License for Providing Telecommunications Services from the supervisory department for the information industry under the State Council on the strength of the Foreign-Invested Enterprise Approval Certificate.

The principal Chinese investor of the foreign-invested telecommunications enterprise shall register the foreign-invested telecommunications enterprise with the administrative department of industry and commerce on the strength of the Foreign-Invested Enterprise Approval Certificate and the Operation License for Providing Telecommunications Services.

Article 18. To provide an international telecommunications service, a foreign-invested telecommunications enterprise must be approved by the supervisory department for the information industry under the State Council and operate through the international telecommunications gateways established with the approval of the supervisory department for the information industry under the State Council.

Article 19. If a foreign-invested telecommunications enterprise violates the provisions of Article 6, the supervisory department for the information industry under the State Council shall order it to correct the mistake within a prescribed timeframe and impose a fine between RMB 100,000 and RMB 500,000. If the correction is not made within the prescribed timeframe, its Operation License for Providing Telecommunications Services shall be revoked by the supervisory department for the information industry under the State Council, and its Foreign-Invested Enterprise Approval Certificate shall be withdrawn by the supervisory department for foreign economic relations and trade which originally issued the same.

Article 20. If a foreign-invested telecommunications enterprise violates the provisions of Article 18, the supervisory department for the information industry under the State Council shall order it to correct the mistake within a prescribed timeframe and impose a fine between RMB 200,000 and RMB 1,000,000. If the correction is not made within the prescribed timeframe, its Operation License for Providing Telecommunications Services shall be revoked by the supervisory department for the information industry under the State Council, and its Foreign-Invested Enterprise Approval Certificate shall be withdrawn by the supervisory department for foreign economic relations and trade which originally issued the same.

Article 21. If approval of an application to establish a foreign-invested telecommunications enterprise is fraudulently obtained by submitting false or forged qualification and certification documents, the approval shall be void, a fine between RMB 200,000 and RMB 1,000,000 shall be imposed and the Operation License for Providing Telecommunications Services shall be revoked by the supervisory department for the information industry under the State Council, and the Foreign-Invested Enterprise Approval Certificate shall be withdrawn by the supervisory department for foreign economic relations and trade which originally issued the same.

Article 22. If a foreign-invested telecommunications enterprise violates the Telecommunications Regulations and other relevant laws and administrative regulations in the provision of telecommunications services, it shall be punished by the relevant authorities according to relevant law.

Article 23.   Listings of domestic telecommunications enterprises overseas must be examined and agreed by the supervisory department for the information industry under the State Council and obtain approval in accordance with the relevant provisions of the State.

Article 24.   For companies and enterprises of the Hong Kong Special Administrative Region, Macau Special Administrative Region and Taiwan Area investing in telecommunications businesses in Mainland China, these Provisions shall apply for reference.

Article 25.   These Provisions shall be implemented as of January 1, 2002.

Appendix C: Telecommunications Regulations of the People's Republic of China[1]

(State Council of the People's Republic of China, September 20, 2000)

Chapter I. General Provisions

Article 1. These regulations are formulated in order to standardize telecommunications market order, maintain the lawful rights and interests of telecommunications customers and telecommunications service operators, protect the security of telecommunications networks and information, and promote the healthy development of the telecommunications industry.

Article 2. Whoever engages in telecommunications activities and telecom-related activities within the territory of the People's Republic of China must abide by these regulations.

---

1. This document is translated with reference to the original version of the Ministry of Information Industry in Chinese and the translated version of China Communications (http://www.telecomm.com) in English. It should be used for general informational purposes and should not be construed as, or used as a substitute for, a legal document.

The term *telecommunications* as mentioned in these regulations refers to activities using wire or wireless electromagnetic systems or photoelectric systems to transport, transmit or receive voice, literature, data, image and any other forms of information.

Article 3. The authority responsible for the information industry under the State Council exercises supervision and administration over the national telecommunications industry in accordance with the provisions of these regulations.

Telecommunications administrative bodies of provinces, autonomous regions and municipalities directly under the central government exercise supervision and administration over the telecommunications industry in their respective administrative regions in accordance with the provisions of these regulations.

Article 4. Telecommunications supervision and management follows the principles of separating government functions from enterprise management, breaking the monopoly, encouraging competition, and promoting development and functioning in a open, fair and impartial way.

Telecommunications services operators shall operate their service in accordance with law, comply with business ethics, and accept supervision and inspection implemented in accordance with law.

Article 5. Telecommunications service operators shall provide telecommunications customers with fast, accurate, secure and convenient telecommunications services at reasonable prices.

Article 6. The security of telecommunications networks is protected by law. No organization or individual shall be allowed to use telecommunications networks to engage in activities that endanger national security, public interests or other persons' lawful rights and interests.

Chapter II. Telecommunications Markets

Section 1 Telecommunications Service License

Article 7. The State applies the licensing system to telecommunications service operation in accordance with telecommunications services classification.

For operating the telecommunications service, a telecommunications operation license issued by the authority responsible for information industry under the State Council or the telecommunications administrative bodies of provinces, autonomous regions or municipalities directly under the central government must be obtained in accordance with these regulations.

No organization or individual shall be allowed to engage in telecommunications service operation activities without obtaining a telecommunications operation license.

Article 8. Telecommunications service is divided into basic telecommunications services and value-added telecommunications services.

Basic telecommunications services refer to services providing public network infrastructure, public data transmission and basic voice communication. Value-added telecommunications services refer to services providing telecommunications and information by using public network infrastructure.

The telecommunications service classification is specified in the "Classified Catalog of Telecommunications Service" attached to these regulations. The authority responsible for the information industry under the State Council may, in the light of actual circumstances, make partial adjustments in the classified items of telecommunications services listed in the catalog and republish them.

Article 9. Operation of basic telecommunications services is subject to approval by the authority responsible for the information industry under the State Council and obtaining the "Operation License for Basic Telecommunications Services".

Operation of value-added services is subject to examination and approval by the authority responsible for the information industry under the State Council and obtaining "Operation License for Trans-regional Value-added Telecommunications Services" if the service covers two or more provinces, autonomous regions or municipalities directly under the central government. And it is subject to examination and approval by the telecommunications administrative body of the province, autonomous region or municipality directly under the central government in question and obtaining an "Operation License for Value-added Telecommunications Services" if the service covers one province, autonomous region or municipality directly under the central government.

Any trial operation of new telecommunications services not listed in the "Classified Catalog of Telecommunications Services" by using new technology should be filed with the telecommunications administrative body of the province, autonomous region or municipality directly under the central government.

Article 10. Anyone who intends to engage in basic telecommunications services shall fulfill the following requirements:

1. The operator is a company specializing in basic telecommunications services established according to law and with state-owned equity or shares not less than 51 percent;
2. Having a feasibility study report and appropriate networking technical schemes;
3. Having financial resources and professional personnel essential to the business activities which it intends to engage in;
4. Having a place and related resources for the service operation;
5. Having a reputation or capability for providing long-term services for customers;
6. Other requirements as specified by the State.

Article 11. Application for operation of basic telecommunications services shall be filed with the authority responsible for the information industry under the State Council, with relevant documents provided under Article 10 of these regulations.

The authority responsible for the information industry under the State Council shall, within 180 days of accepting the application, complete its examination and decide whether to approve or disapprove it. If approved, a "Basic Telecommunications Services License" shall be issued; if disapproved, a written notification shall be sent to the applicant explaining the reason.

Article 12. When examining the application for operating basic telecommunications services, the authority responsible for the information industry under the State Council shall take into consideration the national security, telecommunications network security, sustainable utilization of telecommunications resources, environmental protection, telecommunications market competition status and other factors.

For issuing a "Basic Telecommunications Services License", bidding shall be applied in accordance with relevant provisions of the State.

Article 13. An operator who intends to engage in operating value-added telecommunications services shall fulfill the following requirements:

1. The operator is a company founded according to law;

2. Having financial resources and professional personnel essential to the services it intends to engage in;

3. Having reputation or capability for providing long-term services for customers;

4. Other requirements as specified by the State.

Article 14. Anyone who applies to operate value-added telecommunications services shall file the application with the authority responsible for the information industry under the State Council or with the telecommunications administrative organ of the province, autonomous region or municipality directly under the central government in accordance with the provisions of the second paragraph of Article 9 of these regulations, and submit relevant documents as specified in Article 13 of these regulations.

Where the value-added telecommunications service applied for is subject to examination and approval by relevant competent departments in accordance with provisions of the state, the documents for approval by the competent departments concerned shall also be submitted. The authority responsible for the information industry under the State Council or telecommunications administrative body of the province, autonomous region or municipality directly under the central government shall, within 60 days of receiving the application, complete the examination and decide whether to approve or disapprove it. If approved, the applicant shall be issued a "Trans-regional Value-added Telecommunications Service License" or "Value-added Telecommunications Service License"; if disapproved, a written notice shall be sent to the applicant explaining the reason.

Article 15. Any telecommunications service operator who intends to change the operating subject and scope of service or terminate service operation in the course of operation shall file an application with the authority that has issued the license 90 days in advance, and go through formalities accordingly. In case of ceasing operation, it shall also deal well with the aftermath of operation termination.

Article 16. An operator whose application for operating telecommunications services has been approved shall, with the telecommunications service license obtained according to law, complete registration procedures with the enterprise registration body.

A unit operating a private telecommunications network locally shall file an application in accordance with the requirements and procedures provided in these regulations. It, after approved, must obtain a telecommunications service license, and proceed through registration procedures in accordance with the provisions of the preceding article.

Section 2 Interconnection of Telecommunications Networks

Article 17.   Interconnection and inter working between telecommunications networks shall be effected on the principles of technological feasibility, economic rationality, fairness, impartiality, and coordination.

The leading telecommunications service operator shall not refuse the demand by other telecommunications service operators and private network operators for interconnection and interoperation.

The leading telecommunications service operator as mentioned in the foregoing paragraph refers to the operator who controls the necessary telecommunications infrastructure, and has a larger share in the telecommunications market, is in a position to constitute substantial influence on other telecommunications operators' entry into the telecommunications market.

The authority responsible for the information industry under the State Council shall determine and designate the leading telecommunications operator.

Article 18.   The leading telecommunications operator shall, on the principles of non discrimination and transparency, formulate an interconnection procedure including the process, time limit, list of non binding network elements, etc. The interconnection procedure shall be submitted to the authority responsible for the information industry under the State Council for approval. The interconnection procedure is bonding on the leading telecommunications operator in its interconnection and inter working activities.

Article 19.   For interconnection between public telecommunications networks and that between a public telecommunications network and a private telecommunications network, both parties to the interconnection shall commit to consultation and sign an agreement on the interconnection in accordance with the provisions of the authority responsible for the information industry under the State Council on interconnection administration.

Agreements on network interconnection shall be filed with the authority responsible for information industry under the State Council.

Article 20.   Where both parties to an interconnection fail to reach an agreement on the interconnection, either party may, within 60 days from the date on which one party makes an interconnection request, apply with the authority responsible for the information industry under the State Council or the telecommunications administrative body of the province, autonomous region or municipality directly under the central government for coordination based on the coverage of the interconnection. The body receiving the application shall effect coordination on the principle prescribed in the first paragraph of Article 17 of these regulations to prompt both parties to the interconnection to reach an agreement. Should the coordination fail to lead to an agreement within 45 days from the date on which one party or both parties to the interconnection apply for coordination, the coordinating body shall randomly invite telecommunications technological experts and experts in other related fields to make public demonstration and put forward an interconnection scheme. The coordinating body shall, based on demonstrable conclusion by experts of the effectiveness of the interconnection scheme, make a decision and enforce interconnection terms.

Article 21.   Both parties to an interconnection between networks must effect interconnection and interoperation within the time limit stipulated by the agreement or the decision. Neither party shall terminate the interconnection arbitrarily without the approval by the authority responsible for the information industry under the State Council. Should network-interconnection encounter telecommunications technological barriers, both parties shall immediately take effective measures to remove them. Disputes in connection with interconnection and interoperation shall be settled in accordance with the procedures and methods specified in Article 20 of these regulations by both parties.

In terms of the quality of Interconnection telecommunications should comply with relevant standards laid down by the State. Where the leading telecommunications operator provides network-interconnection for other telecommunications operators, the quality of service should not be lower than that of the same kind of service in its own network and than that of the same kind of service it provides for its subsidiaries or branches.

Article 22.   Settlement and sharing of network-interconnection expenses shall conform to related state stipulations; nothing extra beyond the standard rate shall be charged.

Technical standards, expense settlement procedures, and detailed administrative rules for interconnection between networks shall be worked out by the authority responsible for the information industry under the State Council.

Section 3 Telecommunications Charges

Article 23.   Telecommunications rates must adopt the cost-based pricing principle, with national economic and social development requirements, telecommunications industrial development, telecommunications customers' affordability and other factors taken into consideration at the same time.

Article 24.   Telecommunications charges are divided into market-regulated, government-guided and government-set prices. Value-added telecommunications services apply market-regulated prices or government-guided prices.

For telecommunications services under full market competition, telecommunications rates will apply market-regulated prices.

A Classified Administrative Catalog of Telecommunications Rates applying government-set prices, government-guided prices and market-regulated prices shall be formulated by the authority responsible for the information industry under the State Council after soliciting opinions from the authority in charge of prices under the State Council, and promulgated for implementation.

Article 25.   Government-set telecommunications rates shall be promulgated for implementation with the approval of the State Council following examination by the authority in charge of prices under the State Council.

Margins for telecommunications rates applying government-guided prices shall be formulated and promulgated for implementation by the authority responsible for the information industry under the State Council after soliciting opinions from the authority in charge of prices under the State Council. Telecommunications operators by themselves shall determine the rates within the margins and file them for the record with the telecommunications administrative organ of the province, autonomous region or municipality directly under the central government.

Article 26.   For formulating telecommunications rates applying government-set prices and government-guided prices, hearings shall be held for the expression of opinions from telecommunications operators, telecommunications customers and other interested parties.

Telecommunications operators should, in accordance with the requirement of the authority responsible for the information industry under the State Council and the telecommunications administrative organ of the province, autonomous region or municipality directly under the central government, provide accurate and complete service cost data and other related materials.

Section 4 Telecommunications Resources

Article 27. The State applies unified planning, centralized administration, and rational allocation of telecommunications resources, and institutes a system of paid use.

Telecommunications resources, as mentioned above, refer to limited resources used to realize telecommunications functions like radio frequencies, satellite orbital positions, and telecommunications network numbers.

Article 28. Telecommunications operators should pay telecommunications resources charges for their occupation and use of telecommunications resources. The charging procedures shall be formulated by the authority responsible for the information industry under the State Council in conjunction with the authorities in charge of finance and price under the State Council, and promulgated for implementation after the approval by the State Council.

Article 29. The allocation of telecommunications resources may adopt assignment or auction.

Anyone who obtains the right of use of telecommunications resources should start using the allocated resources and reach the prescribed minimum use scale in the prescribed time limit. No one shall arbitrarily use, transfer or lease telecommunications resources, or change the uses of telecommunications resources without the approval of the authority responsible for the information industry under the State Council or the telecommunications administrative organ of the province, autonomous region or municipality directly under the central government.

Article 30. After a telecommunications resources user obtains telecommunications network code number resources according to law, the leading telecommunications operator and other units concerned are obliged to take necessary technical measures, supporting the telecommunications resource user in realizing the functions of the telecommunications network code number resources.

He shall comply with the provisions otherwise provided, if any, in laws and administrative regulations.

Chapter III. Telecommunications Service

Article 31. Telecommunications operators shall provide services for telecommunications customers in accordance with the standard of telecommunications services prescribed by the State. The type, scope, rate and time limit of services telecommunications operators provide shall be made public and filed with the telecommunications administrative organ of the province, autonomous region or municipality directly under the central government for the record.

Telecommunications customers have the right to choose the use of various telecommunications services opened according to law.

Article 32. If a telecommunications customer applies for installation or reinstallation of telecommunications terminal equipment, the telecommunications operator shall ensure installation and opening of the equipment within the time limit it has announced. If the installation and opening is delayed as a result of any cause or causes on the part of the telecommunications operator, the telecommunications operator shall pay the telecommunications customer a penalty fine in the amount of one hundredth of the installation/reinstallation fee or other fees for each day.

Article 33. In case a telecommunications customer reports a telecommunications service fault, a telecommunications operator shall clear off the fault within a set period—48 hours for cities and towns and 72 hours for rural areas—from the date of receipt of the reporting. In case the fault cannot be cleared off within the time limit, the telecommunications customer be notified timeously, and exempted from the monthly lease fee in the faulty period. However, this is exclusive of telecommunications service faults caused by telecommunications terminal equipment.

Article 34. Telecommunications operators shall provide facilities for telecommunications customers' payments and inquiries. Telecommunications operators shall provide a charging list of domestic long-distance communication, international communication, mobile communication and information services, etc., free of charge if telecommunications customers require.

When abnormally large telecommunications charges accrue to a telecommunications customer, the telecommunications operator shall let the telecommunications customer know as soon as the situation is discovered and take appropriate measures.

Article 35. Telecommunications customers shall pay the telecommunications operator telecommunications charges in time and in full according to the time and way agreed on. In the case of a telecommunications customer failing to pay telecommunications charges in time, the telecommunications operator has the right to require payment of overdue bills, and may ask for an additional defaulting fee of 0.3 percent of the overdue amount for each day.

To the telecommunications customer failing to pay the telecommunications charges that are 30 days overdue, the telecommunications operator may suspend provision of telecommunications services. In a case where the telecommunications customer fails to pay the overdue bills and defaulting fee within 60 days the telecommunications operator may suspend or terminate its service provision, and may press for payment of the overdue bills and defaulting fee according to law.

Operators of mobile telecommunications service may agree upon the time limit and methods of payment of telecommunications charges with telecommunications customers, which are not restricted by the time limit specified in the preceding paragraph.

Telecommunications operators shall restore suspended telecommunications services within 48 hours after the telecommunications customer who delays paying telecommunications charges pays the telecommunications charges in arrears and defaulting fee in full.

Article 36. In case regular telecommunications service is or may probably be influenced by engineering construction, network construction or other causes, the telecommunications operator must notify customers and report to the telecommunications administrative body of the province, autonomous region or municipality directly under the central government within a prescribed time limit.

Where telecommunications service is interrupted due to the causes mentioned in the preceding paragraph, the telecommunications operator shall relieve customers from related charges during the interruption.

In case the telecommunications operator fails to notify the customers timeously of conditions described in the first paragraph of the present Article, it should compensate customers for the loss caused thereby.

Article 37. Telecommunications operators who operate local telephone service and mobile telephone service shall provide customers free of charge with public good telecommunications services such as fire and burglar alarms, emergency medical service, and traffic accident reporting, and ensure that the communication lines are unblocked.

Article 38.   Telecommunications operators shall provide timely access service on an equal and rational basis for group customers who need to be connected to their telecommunications network via a relay line.

Telecommunications operators shall not interrupt access service arbitrarily without approval.

Article 39.   Telecommunications operators should establish an internal service quality control system and may work out, publish and implement an enterprise standard higher than the telecommunications service standard stipulated by the State.

Telecommunications operators shall adopt various methods for soliciting feedback from telecommunications customers, ensuring adequate public supervision, and constantly improving telecommunications service quality.

Article 40.   In the event that service provided by a telecommunications operator falls short of the State set telecommunications service standards (or publicly announced enterprise standards) or telecommunications customers dissent on the payment of telecommunications charges, telecommunications customers have the right to demand satisfaction from the telecommunications operator. In the event of the telecommunications operator refusing solve problems, or the telecommunications customer is dissatisfied with solutions proposed, the telecommunications customer has the right to make complaints to the authority responsible for the information industry under the State Council or the telecommunications administrative body of the province, autonomous region or municipality directly under the central government. The receiver of complaints must handle the case without delay, and reply to the complainant within 30 days from the date on which the complaint is received.

In the event of a telecommunications customer dissenting from the payment of local telephone charges, the telecommunications operator shall also provide the basis for the local telephone call charges at the request of the telecommunications customer, and is obliged to take necessary measures to assist the telecommunications customer in resolving such difficulties.

Article 41.   Telecommunications operators are prohibited from the following acts in the provision of their telecommunications service:

1. Limiting in any form use by telecommunications customers of the service or services offered;

2. Limiting telecommunications customers to buying of telecommunications terminal equipment they appoint, or refusing telecommunications customers' use of their own telecommunications terminal equipment that has obtained network access license;

3. Arbitrarily changing or changing in a disguised form telecommunications rates, increasing or increasing in disguised form charging items in violation of the State stipulations;

4. Refusing, delaying or suspending telecommunications service for telecommunications customers without justifiable reason;

5. Failing to fulfill the commitment publicly made to telecommunications customers or conducting false publicity liable to cause misunderstanding; and

6. Making things difficult for telecommunications customers by devious means, or retaliating against the telecommunications customers who make a complaint.

Article 42. Telecommunications operators are prohibited from the following acts in their telecommunications service operation:

1. Limiting telecommunications customers in their choice of telecommunications services operated by other telecommunications operators according to law;

2. Making unreasonable cross subsidies for various services they operate;

3. Providing telecommunications services at lower-than-cost prices to conduct unfair competition for the purpose of squeezing out competitors.

Article 43. The authority responsible for information industry under the State Council or the telecommunications administrative organ of the province, autonomous region or municipality directly under the central government shall supervise and inspect the telecommunications operators' telecommunications service quality and business activities in accordance with their respective functions and powers. And the result of supervision and random inspection shall be made public.

Article 44. Telecommunications operators must perform their obligation to provide universal telecommunications service in compliance with relevant stipulations of the State.

The authority responsible for information industry under the State Council may determine, by appointing or bidding, a telecommunications operator undertaking the duty of universal telecommunications service.

The Administrative Procedures for Universal Telecommunications Service Cost Compensation shall be formulated by the authority responsible for information industry under the State Council in conjunction with authorities in charge of finance and prices under the State Council, and promulgated for implementation with the approval of the State Council.

Chapter IV. Telecommunications Construction

Section 1 Construction of Telecommunications Facilities

Article 45. The construction of public telecommunications networks, private telecommunications networks and radio and television transmission networks is subject to unified planning and trade administration by the authority responsible for information industry under the State Council.

The construction of any public telecommunications network, private telecommunications network, and radio and television transmission network that belong to a national information network project or a construction project in excess of the quota set by the State is subject to the consent of the authority responsible for the information industry under the State Council before it is submitted to a higher level for examination and approval in accordance the Procedure of Examination and Approval of Capital Construction Projects.

Basic telecommunications construction projects shall be brought into line with the overall construction plans of cities and towns established by people's governments at all levels.

Article 46. Telecommunications facilities shall be installed in parallel with the construction of cities and towns. Telecommunications pipelines and distribution facilities and telecommunications pipelines within the construction area shall be included in the design document of the construction project, and constructed and accepted simultaneously with the construction project. Expenses required shall be included in the budget estimate of the construction project.

When planning or constructing roads, bridges, tunnels or subways, units or departments concerned shall notify in advance the telecommunications administrative body of the province, autonomous region or municipality directly under the central government and telecommunications operators and consult on reservation of telecommunications pipelines and other matters.

Article 47. Basic telecommunications service operators may install additional telecommunications lines and install small antennae, mobile communication base stations and other public telecommunications facilities on civil buildings. But they should notify the owner of title of the building or the user of the building in advance, and pay a use fee to the owner of title of the building rates as stipulated by the people's government of the province, autonomous region or municipality directly under the central government.

Article 48. Where underground, underwater and other concealed telecommunications facilities and high-altitude telecommunications facilities are constructed, such activities must accord with the relevant stipulations of the State.

A basic telecommunications service operator, who intends to construct undersea cables, shall go through necessary formalities according to law after obtaining the consent of the authority responsible for the information industry under the State Council, and solicit opinions from departments concerned. Undersea cables shall be marked on sea charts by the department concerned under the State Council.

Article 49. No units or individual shall be allowed to arbitrarily modify or move telecommunications lines and other telecommunications facilities of others. In case special circumstances make such modification or move essential, consent of the owner of title of the telecommunications facilities in question should be obtained, and the unit or individual requesting the modification or move shall bear the expenses required, and compensate the economic losses caused thereby.

Article 50. Any activities such as construction, production and tree planting, shall not jeopardize the security of telecommunications lines and other telecommunications facilities or hinder the unblocking of lines. When telecommunications security may possibly be imperiled, the telecommunications operator concerned shall be notified in advance, and the unit or individual engaging in such activities shall be responsible for taking necessary protective measures.

Anyone who damages telecommunications lines or other telecommunications facilities or hinders the unblocking of lines in violation of the provisions of the preceding paragraph, should have them recovered and repaired, and compensate for losses caused thereby.

Article 51.   When constructing telecommunications lines, a safe distance shall be kept from telecommunications lines already constructed. Where it is difficult or impossible to avoid passing through, or it is necessary to use the existing telecommunications pipelines, consultation and an agreement shall be made with the owner of title of the telecommunications lines already constructed. In case such consultation fails to lead to an agreement, it shall be solved through consultation by the authority responsible for information industry under the State Council or the telecommunications administrative body of the province, autonomous region or municipality directly under the central government depending on circumstances.

Article 52.   No organization or individual shall prevent or hinder a basic telecommunications service operator from engaging in the construction of telecommunications facilities and providing public telecommunications service from telecommunications customers except in areas where entry is prohibited or restricted by the provisions of the State.

Article 53.   Telecommunications vehicles carrying out tasks of special communications, emergency communications and rush repairs and dealing with emergencies may, with the approval of public security and traffic administrative bodies, not be subject to the restriction of prohibitory signs set for motor vehicles on the premise of ensuring traffic safety and unblocking.

Section 2 Network Access of Telecommunications Equipment

Article 54.   The State shall apply a network access licensing system to telecommunications terminal equipment, radio communication equipment and equipment involved in network-interconnection.

Telecommunications terminal equipment, radio communication equipment and equipment involved in network-interconnection to be connected with a public telecommunications network must comply with the standard set by the State and obtain a network access license.

The list of telecommunications equipment subject to network access licensing control is to be formulated and promulgated for implementation by the authority responsible for the information industry under the State Council in collaboration with the authority in charge of product quality supervision under the State Council.

Article 55. Anyone, who intends to obtain a telecommunications equipment network access license, shall file an application with the authority responsible for the information industry under the State Council. Such application must be accompanied by a test report issued by the telecommunications equipment testing body and approved by the authority in charge of product quality supervision under the State Council, or a product quality certificate issued by the certification body shall be attached.

The authority responsible for the information industry under the State Council shall, within 60 days from the date on which a telecommunications equipment network access license application is received, complete the examination of the application, and the telecommunications equipment test report or product quality certificate. The qualified shall be granted a network access license, and the unqualified given a written reply explaining the reason.

Article 56. Telecommunications equipment manufacturers must ensure quality stability and reliability of the telecommunications equipment with a network access license, and shall not lower product quality and performance.

Telecommunications equipment manufacturers shall stamp a network access license sign on the telecommunications equipment they produce.

The authority in charge of product quality supervision under the State Council shall, in collaboration with the authority responsible for the information industry under the State Council, conduct quality supervision and random inspection, and publish the result of random inspections.

Chapter V. Telecommunications Security

Article 57. No organization or individual shall use telecommunications networks to produce, reproduce, publish and disseminate information containing any of the following content which:

1. Is contrary to cardinal principles defined by the Constitution;
2. May endanger national security, divulge State secrets, subvert state power and sabotage national unification;
3. Injures national honor and interest;
4. Stirs up ethnic hatred, ethnic discrimination, and sabotages national unity;

5. Sabotages the policy of the State on religion, publicizes heretical cults and feudal and superstitions;

6. Spreads rumors to disrupt public order and undermine social stability;

7. Spreads pornography, salacious material, encourages gambling, violence, murder, terror or instigates crime;

8. Insults or slanders other persons, and infringes lawful rights and interests of other persons; and

9. Is prohibited by relevant laws and administrative regulations.

Article 58. Any organization or individual is prohibited from any of the following acts that may endanger telecommunications network security and information security:

1. Deleting or modifying stored, processed and transmitted data and application software of a telecommunications network;

2. Using a telecommunications network to engage in activities of stealing or destroying other person's information and injuring other person's lawful rights and interests;

3. Intentionally producing, reproducing and propagating computer viruses or attacking other persons' telecommunications networks or other telecommunications facilities in other ways;

4. Other acts that may do damage to telecommunications network security and information security.

Article 59. No organization or individual is allowed to undertake any of the following acts that may disrupt telecommunications market order:

1. Arbitrarily operating international, or Hong Kong Special Administrative regional, Macao Special Administrative regional and Taiwan regional telecommunications services by leasing international telecommunications private lines, illicitly setting up relay equipment or adopting other means;

2. Unlawfully connecting to other persons' telecommunications lines, reproducing other persons' telecommunications code numbers, and knowingly using unlawfully connected telecommunications facilities and reproduced code numbers;

3. Forging or distorting telephone cards and other valued certificates of telecommunications service;
4. Proceeding through network access formalities and using a mobile phone with a false or other person's I.D. card.

Article 60.   Telecommunications operators shall establish a comprehensive set of rules and regulations, and apply a security assurance responsibility system in compliance with the provisions of the State concerning telecommunications security.

Article 61.   In their telecommunications network design, construction and operation, telecommunications operators shall endeavor to achieve synchronized planning construction and operations for national security and telecommunications network security.

Article 62.   A telecommunications operator, who deems information transmitted on the telecommunications network is subject to prohibitions as listed in Article 57 of these regulations in its public information service, shall immediately cease transmission, preserve related records and report the matter to relevant State bodies.

Article 63.   Telecommunications customers shall be held responsible for the content of information they use the telecommunications network to transmit and the consequences thereof.
    Where the information a telecommunications customer intends to transmit involves State secrets, security measures must be taken in compliance with the provisions of the Law on Guarding State Secrets.

Article 64.   In case of major natural calamities or other emergencies, the authority responsible for the information industry under the State Council may dispatch (with the approval of the State Council) all telecommunications facilities to ensure important communications are unblocked.

Article 65.   Any international communication service within the People's Republic of China must be conducted via the international communication gateway set up with the approval of the authority responsible for the information industry under the State Council.
    For communications between the Chinese inland and the Hong Kong Special Administrative region, the Macau Special Administrative region and

Taiwan region, the provisions of the preceding paragraph shall be complied with.

Article 66.   The freedom of telecommunications customers in using the telecommunications service is protected by law. No organization or individual shall be allowed to inspect telecommunications content for any reason, except inspection of that telecommunications content by the public security body, national security body or people's procuratorate in accordance with the procedures provided by law for the sake of national security or for the need to investigate a criminal offense.

Chapter VI. Penalties

Article 67.   Whoever acts in violation of the provisions of Articles 57 and 58 of these regulations shall, if such acts constitute a criminal offense, be subject to criminal prosecutions pursuant to law. If such acts do not constitute a criminal offence, penalties shall be imposed thereon by the public security body and national security body in accordance with the provisions of relevant laws and administrative regulations.

Article 68.   Whoever commits one of the acts listed in Items (2), (3) and (4) of Article 59 of these regulations to disturb telecommunications market order shall, if such acts constitute a criminal offence, be subject to criminal prosecution. If such acts do not constitute criminal offences, the authority responsible for the information industry under the State Council or the telecommunications administrative body of the province, autonomous region or municipality directly under the central government shall, according to their respective functions and powers, order it to make corrections, confiscate illegal gains, if any, and impose a fine of not less than 3 times but not more than 5 times the illegal gains. If no illegal gains are involved or the illegal earnings are less than 10,000 yuan, a fine of not less than 10,000 yuan but not more than 100,000 yuan shall be imposed.

Article 69.   Whoever forges, illegally uses and transfers telecommunications operation licenses and telecommunications equipment network access licenses or fabricates network access license numbers on telecommunications equipment shall be subject to the confiscation of illegal gains, and be subject additionally to the imposition of a fine 3–5 times the amount of illegal gains by the authority responsible for information industry under the State

Council or the telecommunications administrative body of the province, autonomous region or municipality directly under the central government according to their respective functions and powers. Where no illegal gain is involved or the illegal gain is less than 10,000 yuan, a fine in the range of 10,000 to 100,000 yuan shall be imposed.

Article 70. Whoever commits one of the following acts in violation of the provisions of these regulations shall be ordered to make corrections, be subject to the confiscation of illegal gains, and suffer a fine 3 to 5 times the amount of illegal gain by the authority responsible for information industry under the State Council or the telecommunications administrative body of the province, autonomous region or municipality directly under the central government according to their respective functions and powers. Where there is no illegal gain or the illegal gain is less than 50,000 yuan, a fine in the range of 100,000 to 1,000,000 yuan shall be imposed; if the circumstances are serious, suspension of operation for rectification of the harm shall be ordered:

1. Violating the provisions of the third paragraph of Article 7 or having actions listed by Item 1 of Article 59 of these regulations to operate a telecommunications service without approval or beyond the approved service scope;
2. Setting up an international communication gateway to conduct international communications without the approval by the authority responsible for the information industry under the State Council;
3. Arbitrarily using, transferring and leasing telecommunications resources or changing the uses of telecommunications resources;
4. Arbitrarily suspending interconnection of networks or access service; Refusing to perform universal service obligations.

Article 71. Whoever commits one of the following acts in violation of the provisions of these regulations shall be ordered to make corrections and suffer the confiscation of illegal gains, and an imposed fine of not more than three times the amount of the illegal gains by the authority responsible for information industry under the State Council or the telecommunications administrative body of the province, autonomous region or municipality directly under the central government according to their respective functions and powers. Where there is no illegal gain or the illegal gain is less than

10,000 yuan, a fine in the range of 10,000 to 100,000 yuan shall be imposed; if the circumstances are serious, suspension of operation for rectification shall be ordered:

1. Charging extra fees in violation of stipulations in the interconnection and interoperation between telecommunications networks;
2. In case of inter network communication technical barriers, failing to take effective measures to remove;
3. Arbitrarily providing other persons with the content of information telecommunications customers use telecommunications network to transmit;
4. Refusing to pay the use fee of telecommunications resources according to stipulations.

Article 72.   Whoever conducts unfair competition in telecommunications service operations in violation of the provisions of Article 42 of these regulations shall be ordered to make corrections, and be penalized 100,000 to 1,000,000 yuan by the authority responsible for information industry under the State Council or the telecommunications administrative body of the province, autonomous region or municipality directly under the central government within their respective responsibilities. If the circumstances are serious, suspension of operation for rectification shall be ordered.

Article 73.   Whoever commits one of the following acts in violation of the provisions of these regulations shall be ordered to make corrections and suffer an imposed fine in the range of 50,000 to 500,000 yuan by the authority responsible for information industry under the State Council or the telecommunications administrative body of the province, autonomous region or municipality directly under the central government within their respective responsibilities; if the circumstances are serious, suspension of operation for rectification shall be ordered:

1. Refusing other telecommunications operators' request for interconnection;
2. Refusing to implement the decision on interconnection and inter working made by the authority responsible for information industry under the State Council or the telecommunications administrative organ of the province, autonomous region or municipality directly under the central government according to law;

3. Providing other telecommunications operators with inter working between networks whose quality of service is lower than that of its own network and its subsidiaries and branches.

Article 74. Where a telecommunications operator, in violation of the provisions of the first paragraph of Article 34 and the second paragraph of Article 40 of these regulations, refuses to provide free of charge telecommunications customers with a charging list for domestic long-distance communications, international communications, mobile communications, information service, etc., or refuses to provide free of charge telecommunications customers with the basis for local telephone call charges when telecommunications customers dissent from the payment of local telephone call charges and request review of such charges, the telecommunications administrative body of the province, autonomous region or municipality directly under the central government shall order it to make corrections, and to make an apology to the telecommunications customer in question; those who refuse to make corrections and make an apology, shall be given a warning, and have imposed a fine of 5,000 to 50,000 yuan.

Article 75. Whoever acts in violation of the provisions of Article 41 of these regulations shall be ordered to make corrections, to make an apology to telecommunications customers, and compensate telecommunications customers' losses. To those who refuse to make corrections, to make an apology and to compensate telecommunications customer's loses, a warning shall be given, and a fine imposed of 10,000 to 100,000 yuan. If the circumstances are serious, suspension of operation for rectification shall be ordered.

Article 76. Whoever commits one of the following acts in violation of the provisions of these regulations, the telecommunications administrative body of the province, autonomous region or municipality directly under the central government shall order it to make corrections, and impose a fine of 1,000 to 100,000 yuan:

1. Selling telecommunications terminal equipment without network access license;

2. Unlawfully preventing or hindering telecommunications operators from providing public telecommunications services for telecommunications customers;

3. Arbitrarily changing or moving other person's telecommunications lines and other telecommunications facilities.

Article 77. Anyone who lowers product quality and performance after obtaining network access license for telecommunications equipment in violation of the provisions of these regulations shall be penalized by authorities in charge of product quality supervision in compliance with the provisions of relevant laws and administrative regulations.

Article 78. To penalize anyone who has committed an act prohibited by Articles 57, 58 and 59 of these regulations, and if the circumstances are deemed sufficiently serious, the issuer shall revoke the telecommunications operation license. The authority responsible for the information industry under the State Council or the telecommunications administrative body of the province, autonomous region or municipality directly under the central government shall notify the authority in charge of business registration after revoking a telecommunications operation license.

Article 79. Any functionary of the authority responsible for the information industry under the State Council or telecommunications administrative bodies of provinces, autonomous regions or municipalities directly under the central government who commits such acts as neglect of duty, abuse of power, and malpractice for selfish ends, which constitute criminal offenses, shall be subject to criminal prosecution pursuant to law; where such acts do not constitute criminal offenses, administrative sanctions shall apply.

Chapter VII. Supplementary Provisions

Article 80. Detailed rules for investment in and operation of telecommunications service in the People's Republic of China by foreign organizations or individuals and for investment in and operation of telecommunications service in the Chinese inland by organizations or individuals from the Hong Kong Special Administrative Region, Macao Special Administrative Region and Taiwan region, are to be formulated separately by the State Council.

Article 81. These regulations come into effect as of the date of promulgation.

Appendix D: A Classified List of Telecommunications Services

I. Basic Telecommunications Services

1. Fixed network domestic long-distance and local telephone services;
2. Mobile network telephone and data services;
3. Satellite communication and satellite mobile communication services;
4. Internet and other public data transmission services;
5. Leasing and selling services of bandwidth, wavelength, optical fiber, optical cable, pipeline and other network elements;
6. Network bearer and access and network external contracting services;
7. International telecommunications infrastructure and international telecommunications services;
8. Radio paging service;
9. Resale basic telecommunications service.

Services listed in 8 and 9 are subject to management in the light of value-added telecommunications services.

II. Value-Added Telecommunications Services

1. Electronic mail;
2. Voice mailbox;
3. Online information base store and retrieval;
4. Electronic data interchange (EDI);
5. Online data processing and transaction processing;
6. Value-added fax;
7. Internet access service;
8. Internet information service;
9. Videoconferencing service.

List of Abbreviations

ADSL	asymmetric digital subscriber line
ARPU	average revenue per user
MOU	minutes of usage
BT	British Telecom
CAD	computer aided design
CAS	Chinese Academy of Sciences
CCCCP	Central Committee of the Chinese Communist Party
CCF	China-China-Foreign
CDMA	code division multiple access
CEPA	Closer Economic Partnership Arrangement
CEWC	Central Enterprise Working Commission
CHGR	compound half-yearly growth rate
China Unicom	China United Telecommunications Corporation
CITIC	China International Trust and Investment Corporation
CPE	customer premises equipment

CVC	cost-variance coefficient
DGP	Directorate General of Posts
DGT	Directorate General of Telecommunications
DTA	Department of Telecommunications Administration
EDI	Electronic data interchange
EEACT	Eastern Extension Australia and China Telegraph Company
EOR	enterprise-owned revenue
FCC	Federal Communications Commission
GNTC	Great Northern Telegraph Company
GPRS	General Packet Radio Service
GPS	Global Positioning System
GTE	General Telephone and Electronic Corporation
GW	gateway
HKT	Hong Kong Telecom
IAI	initial address message with additional information
IAM	initial address message
IDD	international direct dial
IP	Internet Protocol
IPR	intellectual property rights
ITU	International Telecommunication Union
LAN	local area network
MEI	Ministry of Electronic Industry
MEP	Ministry of Electrical Power
MFN	most-favored-nation
MII	Ministry of Information Industry

MISC	Mobile Information Service Center
MOR	Ministry of Railways
MPT	Ministry of Posts and Telecommunications
MRFT	Ministry of Radio, Film, and Television
MSCs	mobile switching centers
PHS	Personal Handyphone System
PLA	People's Liberation Army
POTS	plain old telephone service
PTAs	Posts and Telecommunications Administrations
PTIC	Posts and Telecommunications Industry Corporation
SAIC	State Administration for Industry and Commerce
SARFT	State Administration of Radio, Film, and Television
SCSCNII	State Council Steering Committee of National Information Infrastructure
SMS	short message service
SOE	state-owned enterprise
SPC	State Planning Commission
SPC	stored program control
TD-SCDMA	time division synchronous code division multiple access
TM	tandem
TS	toll switch
USO	universal service obligation
VoIP	Voice over IP
VSAT	very small aperture terminal
WAP	wireless application protocol
WRC	World Radio Conference

WTO   World Trade Organization

ZTE   Zhongxing Corporation

About the Authors

Xu Yan is assistant professor of information systems management at the Hong Kong University of Science and Technology. His research and teaching interests include information systems and telecommunications policy.

In 1997, he received his Ph.D. from the Strathclyde Business School, Glasgow, Scotland. He has worked at the Beijing University of Posts and Telecommunications, the Ministry of Posts and Telecommunications in China, British Telecom Laboratories, and the International Telecommunications Union (ITU). He has undertaken research and provided consulting and executive training for government agencies and leading world organizations, such as the ITU, China's Ministry of Information Industry, the Pakistan Telecommunications Authority, British Telecom, China Mobile, China Telecom, Elcoteq, Nokia, and Nortel. He was recently appointed by the Audit Commission of the Hong Kong government as a consultant charged with the review of Hong Kong's telecommunications policy.

Xu Yan is currently on the board of directors of the International Telecommunications Society. He is also member of the IT Manpower Taskforce of the Hong Kong government. He has published extensively in internationally renowned journals, such as *Telecommunications Policy, Communications and Strategies, IEEE Communications,* and *Info.*

Douglas Pitt was formerly at Strathclyde University in Glasgow, Scotland, where he occupied the chair of organizational analysis and was dean of the Strathclyde Business School. Since 2001, he has been based in Cape Town,

South Africa, where he is a professor and dean of the Faculty of Commerce at the University of Cape Town.

Mr. Pitt has been involved with issues in the telecommunications sector since completing his Ph.D. in the late 1970s at the University of Manchester, England, on organizational transformation in the British Post Office. He has worked on deregulatory developments in Britain and the United States and is widely published. He has also acted as a consultant to British Telecom. He is currently chief editor of the journal *Telecommunications Policy* and has worked on issues concerning Chinese deregulatory developments for the past decade with Xu Yan.

Index

Telecommunications Deregulation and the Information Economy, Second Edition, James K. Shaw

Telemetry Systems Engineering, Frank Carden, Russell Jedlicka, and Robert Henry

Telephone Switching Systems, Richard A. Thompson

Understanding Modern Telecommunications and the Information Superhighway, John G. Nellist, and Elliott M. Gilbert

Understanding Networking Technology: Concepts, Terms, and Trends, Second Edition, Mark Norris

Videoconferencing and Videotelephony: Technology and Standards, Second Edition, Richard Schaphorst

Visual Telephony, Edward A. Daly and Kathleen J. Hansell

Wide-Area Data Network Performance Engineering, Robert G. Cole and Ravi Ramaswamy

Winning Telco Customers Using Marketing Databases, Rob Mattison

World-Class Telecommunications Service Development, Ellen P. Ward

For further information on these and other Artech House titles, including previously considered out-of-print books now available through our In-Print-Forever® (IPF®) program, contact:

Artech House	Artech House
685 Canton Street	46 Gillingham Street
Norwood, MA 02062	London SW1V 1AH UK
Phone: 781-769-9750	Phone: +44 (0)20 7596-8750
Fax: 781-769-6334	Fax: +44 (0)20 7630-0166
e-mail: artech@artechhouse.com	e-mail: artech-uk@artechhouse.com

Find us on the World Wide Web at:
www.artechhouse.com